FURTIVE FAUNA

A field guide to the creatures who live on you

Roger M. Knutson

Ten Speed Press
Berkeley, California

Ten Speed Press
P.O. Box 7123
Berkeley, California 94707

Distributed in Australia by E. J. Dwyer Pty. Ltd., in Canada by Publishers Group West, in New Zealand by Tandem Press, in South Africa by Real Books, in Singapore and Malaysia by Berkeley Books, and in the United Kingdom and Europe by Airlift Books.

Cover and interior design by Fifth Street Design

Illustrations by Joe Judd

Library of Congress Cataloging-in-Publication Data

Knutson, Roger M., 1933–
Furtive fauna : a field guide to the creatures who live on you / Roger M. Knutson
p. cm.
Originally published: New York, N.Y., U.S.A: Penquin Books, 1992.
Includes bibliographical references.
ISBN 0-89815-827-3
1. Body, Human—Microbiology—Popular works. 2. Symbiosis—Popular works. I. Title.
QR171.A1K68 1996
616.9'6—dc20 96–17234
CIP

First Ten Speed printing, 1996

Printed in Canada

1 2 3 4 5 6 7 8 9 10 — 00 99 98 97 96

Dedicated to Melvin Knutson, whose thirst for knowledge changed us all

••••••••••

Thanks are due to the reference librarians at Luther College and to Jeanne K. Hanson, a great auctioneer.

Contents

Chapter 3
NEIGHBORS

Chapter 4

RESIDENTS

Chapter 1

THE FURTIVE FAUNA'S POINT OF VIEW

Why Anything Would Want to Live on Us and How Our Relationship Came to Be

Under the pressure of completing an examination, a student defined a parasite as "something that lives right on the thing it lives off of." More formally, a parasite derives some or all of its nourishment from its host. Host—that does not sound like a problem; most of us enjoy hosting, at least in the usual sense. Breaking out the bourbon and calling for the musicians isn't the hardest thing one has to do. As host to a parasite we have different obligations: not just providing food but becoming food. Becoming food and shelter for a creature that many of us wouldn't even like to look at isn't most people's idea of a party. But such a close relationship is not all that uncommon. Just about every animal has its intimate companions, its parasites. Surely people don't have such disgusting relationships, do they? Of course we do, and we have for as long as there have been people. Linguists have suggested that one of the very first words in the human vocabulary over 100,000 years ago was a term for flea or tick. And today we house a wonderful array of beasties, with relatively few discomforts and only an occasional outbreak of the plague to remind us that we are more a community than an individual.

What sort of critters live on and around us? Most are just about too small to see, or, if they are visible, they move quickly. Most are at least a bit secretive, and some live in places we may not even want to look, or can't see without the help of extremely cooperative friends. They range from the almost totally innocuous to the frighteningly intrusive. And they are all interesting. To know a person well you should know his or her friends and close companions. To know *yourself* well, you should know your closest and most enduring, if not endearing, companions: your parasites. These parasites are not like your in-laws; they are smaller and easier to maintain.

Parasites come in at least two types, based on just how intimate their relationship with you is. Endoparasites live on the inside, genuinely out of sight if not out of mind (or body). As a group, they are disgusting creatures with no redeeming characteristics; they devour us from within and by these depredations weaken and sicken us. We will mention them no further. Ectoparasites live on or visit our outer surfaces and, by contrast with those on the inside, are of an almost sunny disposition. They have served over the centuries to enlighten, amuse, even educate us.

Ectoparasites, our friends, have brought about changes in humankind through the ages, and we have brought about changes in them. We have changed the ectoparasites into better hiders, better jumpers, and faster feeders. It is for survival, not entertainment, that fleas are able to jump as far and as quickly as they can. The lice that hide in places we are reluctant to look hold on as well as they do because any of their potential ancestors who did not hold on were left behind during some prehuman's dash across the savanna. Any bedbug who tried to feed on a wide-awake human was surely discovered and slapped into nonexistence. Only those who approached us during sleep lived to have offspring. We have helped our parasites become better at finding us, holding on to us, and feeding from us by eliminating those we could catch or avoid; surely a noble effort on our part.

And have these relatively pleasant companions changed us? In our unrelenting and so far unsuccessful efforts to rid ourselves of invaders, we have been selected over thousands of generations for our capacity for close vision. Those who could see fleas or lice or ticks more easily were better at grooming the skin, hair, and surfaces of their companions, as all primates do. Good grooming and reproductive success are commonly related to one another. We were probably prepared for reading and writing later in human history by those early efforts to see small dark creatures against the skin of others. And there was certainly natural selection among those early humans for enhanced powers of persuasion. Someone who can talk another person into looking them over for tiny ectoparasites would have no trouble selling Veg-o-matics to the unwary. And one person who could convince another to remove those crawling or grasping things from out-of-the-way parts of their anatomy would have a better chance of living well and reproducing abundantly with a lower load of parasites. The habits and skills that make a good used-car salesman probably go as far back into human history as those habits and skills that make a good librarian.

In general, only those ectoparasites able to avoid our attention, to feed quickly or remain invisible, left any progeny in future furtive generations, and those of the human persuasion better able to catch them by hand or mouth had more time for their own reproduction. That we have been almost but not always able to catch the little creatures before they finish their lunch speaks of the evenness of the contest. Some of the quickest of us may be able to catch a fly on the wing, with or without chopsticks, but most of us catch only a few. They are just as good at getting away as we are at catching them.

Some of the creatures described in this book have been with us for millions of years, both watching and participating in human evolution. Others we have only recently acquired, as our habits of dress and housing have changed. Some are friendlier than others. Some visit us only infrequently, when they need something, like the unwelcome relatives who only turn up when they need a loan for a down payment. Some are close neighbors who feel free to drop in when they need a cup of sugar or a snack. Others are with us daily and participate in our most intimate moments. A few are permanent residents for whom our skin is their whole world; a place of varying climates, forests of hair, and exposed clearings. Each of the permanent residents lives with the ever-present danger that their world will one day rise up and smite them. A considerable number are too small to see at all without some sort of artificial magnification. You probably believe they are not there. Trust me, they are. Each of us is a whole ecological landscape for these tiny ones, with every pore both a potential hiding place and a potential death trap; with every hair both a source of shade and something to hold on to when that unpredictable flood of sweat threatens to wash everything small away.

Collectively, our friendly and not-so-friendly ectoparasites have brought about the winning and losing of more battles than Napoleon and have done more for population control than Margaret Sanger. We often speak unkindly about them, calling them things like "vermin." We have written songs and poems about them that both praise and damn. Robert Burns, the Scottish poet, wrote of the louse with real affection in a poem titled, "To a Louse, on seeing one on a lady's bonnet at church." A few of the ectoparasites are actually endangered species, not because of our neglect, but because of our direct and continuing efforts to eliminate them from the world. They are creatures in whose favor very little will be said by most, but they are interesting and at least potentially significant for human welfare and human understanding.

The things that live on and very close to us deserve

another look if for no other reason than self-knowledge. If we do not understand their history, we may be condemned to relive it. Time was when everyone harbored enough parasites to make them the common subject of poetry and art, and a source of common phrases such as "a flea in one's ear." They are still close by in our environment, waiting for us to get careless about public health or private cleanliness; ready to move back to their free room and board.

The Human Landscape

Viewed from the perspective of the fauna likely to inhabit us, the surface of the human organism must seem like a land of milk and honey, or whatever the equivalent foods would be for the home-seeking louse, flea, or mite. The rest of the world is by comparison cold, windy, and mostly without food. "Find yourself one of these," the mother mite must say when admonishing the children about how to get along in the world. She might add:

> There is plenty of food. Good nutritious flakes of skin are coming loose constantly, tasty fluids ooze out of holes like water from a spring, and there are good places to hide if you need to. There are small holes all over the surface, and there are tall straight stalks to hold onto. It's warm almost all the time, and warmer in some places than in others. There are no bad animals anywhere on the surface, and the whole landscape is quiet for at least part of every day. You couldn't ask for a better place to live or raise a family. But there is one problem. Now and then your home will spend a lot of time looking itself over for anything foreign, and it thinks you are foreign. The real tiny ones among us don't have any trouble because they are too small to be seen, but those of us who are larger have to be very fast or very secretive, or our lives are really up for grabs.

Mother fleas or lice probably give much the same advice.

The place an animal lives is called its habitat. While

some ectoparasites just visit, for those that live right on us we are the habitat. The characteristics of any habitat include temperature, substrate (what's underneath), moisture availability, and places to hide. As habitat for small creatures, we provide an extremely varied set of locations in what to us is a small space. To our residents and visitors, though, it is the whole world.

Following poet Robert Burns's advice to see ourselves as others see us, we can look at ourselves as a landscape. To the common furtive resident or visitor a human body of ordinary size is about what a square mile would be to us, but a far more interesting square mile than any of us is likely to encounter in the full-sized world. There is a tropical climate with a most remarkable array of steep hills and deep valleys, some providing abundant cover and others almost desertlike in their barrenness. Some places—such as the face and hands—are in almost constant motion, while others move only during vigorous exercise. On much of the surface, sudden floods of sweat will wash away almost anything that isn't fastened down, and other places such as the inside of the ear cavity almost never see liquid of any sort. In some places, the hairs are close together and long, while in others they are short and separated by distances that are impressive to a creature as small as a pinhead.

Add clothing to the body and whole new habitats become available. Places that were merely warm become oppressive, and places that were hardly warm enough for survival become comfortable. Humidity on the surface goes up dramatically. When people don't wear shoes there is no athlete's foot; the fungus that causes it can't live in such a dry, airy place. Body lice found their present home only when we started wearing coverings of some sort, and now they live in our clothes and only visit the surface of the skin when they are hungry. If we are a wonder to ourselves, imagine what we must be to one of our tiny residents wandering over that almost-limitless surface, stopping in this pore or that, resting in the temporary shade of one or another of those hairs, sipping from one of the fountains or nibbling just a flake or two of loose skin cells. No wonder we are hosts to so many: we are prime real estate.

Disease and the Furtive Fauna

Other than a kind of natural revulsion at being food for some other creature, the major reason that most people object to the presence of ectoparasites on or around themselves is fear of the diseases they might carry. But actually these creatures are not malevolent; they are just trying to make a living in any way they can, no different from the rest of us. Most of them don't carry diseases most of the time, and when they do, they often die of it before they can transmit it to us. It is also true that most of the diseases they carry and transmit to us they acquired from humans in the first place. Nearly all of the ectoparasite-borne diseases—plague, typhus, and several louse-borne fevers—are not present in furtive fauna or in us until there are destructive situations produced by human activity. War, famine, and revolution can hardly be blamed on even the most intelligent fleas or lice, but these two groups of insects have been blamed for almost every serious outbreak of ectoparasite-borne disease.

Everyone has read, of course, about the horrors of bubonic plague, and probably knows that it is carried by fleas. But on closer examination it becomes hard to blame the fleas. Plague kills fleas just as surely as it kills rats or people. The spread of the disease requires that there are plague-infected rats, that the rats have fleas, and that those fleas are willing to feed on humans. Most rat fleas are more discriminating than that and prefer rats. Most human fleas are reluctant to give up their relationship with us and spend any time on rats. Under the worst possible set of circumstances, plague, which is normally present in a few rats, increases dramatically in the rat population either because there are too many rats or because of particularly effective fleas. When the rats die of the disease and their bodies cool, their fleas leave them, and when the fleas begin to starve they are willing to consider humans as an alternative. The fleas are also sick by this time and are not fully responsible for their actions. If there are enough rats close to enough people, some people will become infected and can then transmit the plague to their own fleas. If the human fleas are present in large enough numbers, they become infected and can carry the disease from one person to the next. If the human population is sufficiently dense, and if most of them have fleas, a serious epidemic might result. All the ifs in the previous sentence suggest why epidemics of plague are now rare. Plague is still present in the world as a disease of rodents, and commonly kills rodents in the western United States. But it causes few problems for us as long as large numbers of rodents and fleas are not close to us.

Consider this, however. The best carrier of bubonic plague from rats to humans is the tropical rat flea. Not a problem for us if we don't live in the tropics, right? Wrong! The tropical rat flea is found as far north as Canada. Humans have created abundant tropical habitats for the rats and their fleas all over the world. Not just homes, but gigantic enclosed malls and domed sports arenas can be habitats for both. The malls and domes provide outstanding food and housing for rats if not carefully managed, and enormous numbers of people gather there and then scatter again. A visitor to a superbowl in Minneapolis can be anywhere in the world within twenty-four hours. Concern for the control of rats, fleas, and people in such public locations should have a high place on our agenda of public-health considerations. As long as rats and fleas don't develop resistance to the materials we use to control them, we have nothing to worry about. On second thought, maybe we should worry a little bit. In any case, it is clear that blaming the fleas or the lice for our diseases is like killing the messenger. We, not our tiny visitors, bear responsibility for epidemics of bubonic plague or typhus.

Chapter 2

VISITORS

Creatures for whom we are merely an occasional stopping place

Nearly any creature that flies or crawls and is small enough to pause on us without our knowing it could fit in the visitor category. Most are no more troublesome than a cloud of gnats that swarms around your head or that only want to sop up a bit of our sweat or the fluid from our tears. A few are worth recognizing for their inherent interest or because some of them might cause us trouble—just as it would be important to recognize the difference between a mob of black-jacketed bikers parked in front of your house and one well-dressed person carrying a vacuum cleaner. Both might mean trouble, but one is more likely to cause panic than the other.

The visitors we will be dealing with all use some part of us for food, and a few of them use quite a lot of us for food. Some of them are likely to cause disease or damage, and we aren't usually well adapted to conditions caused by what are only occasional visitors. The diseases that give us the least trouble are those we have lived with and adjusted to for hundreds of generations. Long-term survival for a disease producer requires that it not cause us too much trouble. Killing the goose that lays golden eggs has never been a good idea for any sort of creature, and for the disease producer, we are the goose. Any parasite benefits greatly from keeping its host alive, if not completely healthy.

Ticks *(Ixodidae)*

There are no ticks with especially close or dependent relationships with people. Hurrah, you are saying. Don't cheer too soon. Many kinds of ticks are perfectly willing to use us for food if they find us, and many of them can lie in wait for years, hoping something edible wanders by close enough for them to climb aboard. And since whatever climbs aboard you most probably got its meal from something not human, and since ticks are capable of carrying, supporting, and injecting into you more kinds of disease-causing microbes than almost any of our other visitors, there are good reasons for recognizing them.

Ticks are not insects but a close relative of mites and spiders. They have eight legs and a flat, hard body. If you don't care to count legs, their habit of motion will almost surely identify them. They scuttle over your surface. It's a good idea to find them while they are still scuttling. If something nearly round, dark, about the thickness of cereal box cardboard and with more legs than you can easily count is scuttling over any part of you, that is sufficient evidence for you to shudder and shout, "It's a tick, it's a tick!" You don't have to know its name to address the problem. If you pick it up and cannot crush it between thumb and forefinger, then it almost certainly is a tick. Most all of the commonly encountered ticks are hard, thin, dark, and scuttling.

Like many of their arthropod relatives, ticks hatch from eggs and grow through three distinct stages: larva (too small to see), nymph (just visible), and adult (ready to copulate and lay eggs). You are most likely to see and respond to the adults. Ticks at any of the stages of development can use us for food, and at each stage they need what those who study them casually call a "blood meal" before they can grow to the next stage. With the blood, they pick up any microbes that might be circulating in their temporary host and carry them to their next meal, thus spreading disease. Ticks, like crabs and lobsters, have a hard covering on their outside that must be periodically shed if they are to grow, so the life cycle for a tick is something like hatch, eat, shed, grow, find a new host, eat, shed, grow, find a new host, eat, copulate, lay eggs, die.

The biggest problem ticks have is finding an appropriate host for their next meal. They cannot fly, they crawl only slowly, and generally cannot travel without some help for more than a few yards from where they were hatched. Lacking the machinery for movement through space, ticks have become masters at moving through time. When they are ready for their next meal they can wait for months, years, or even decades for the right host to come along. This is called questing behavior. They crawl or scuttle up on a blade of grass or a twig and settle in. If nothing with blood walks by they wait, and wait, and wait. How do they know when something with blood is walking by? They don't see well, but they have an extremely sensitive and rapid response to a whiff of carbon dioxide, which all animals exhale, or to the faint odor of butyric acid, exuded from the skin of many animals. Given such a whiff, the dormant tick is instantly activated, even though it may not have moved for years. It scuttles rapidly or almost leaps onto anything that brushes by. Since ticks lay large numbers of eggs and the offspring don't often travel far, it is possible to walk through a whole batch of activated ticks, or, heaven help you, lie down among them.

Whenever you have been out in nature, especially if you have spent any time lying on the ground, it's worth having a close friend look you over for ticks. Assure your friend that all primates groom each other. The tick, or ticks if you are very unlucky or very warm (they seem to be attracted to the more hot-blooded types), should still be scuttling around on you. Should you lack friends willing to look for ticks right away, the tick or ticks might have settled down in one place before you find them. What they do is bury the front part of their head, called a hypostome (translation: below-mouth), in your skin and begin feeding. It sounds as if it should hurt, but ticks don't want to be found so are able to anesthetize the area before they begin feeding. You don't feel a thing. The tick, having attached an important part of its head to you, is understandably reluctant to leave. There are almost as many ways of getting rid of ticks as there are ticks, but most of them don't work or work poorly. Applying a hot match to the tick might work if the tick were on some totally nonhairy part of your landscape, but since ticks often prefer shadier locations, fire is probably out. The best you could hope for would be that a hot match might kill the tick, but it would still be firmly attached to your skin. Covering the tick with nail polish or oil is often suggested, and such treatment will eventually kill the tick, but remember you are dealing with a creature that hasn't eaten for maybe ten years and could probably wait another ten. You likely don't have that much time to wait. Telling the tick to leave won't work either; ticks hear even less well than they see. Besides, the tick's hypostome is attached to you by a sort of salivary epoxy that makes it difficult to detach itself quickly even if it wanted to. Removal is probably best done with a pair of tweezers small enough to grasp only the head of the tick. Pull the head and attached tick out slowly and apply a dab of antiseptic.

An especially macho response to the presence of a tick is to let it feed and grow. They can become ten times their initial size in a few days. Have fun, compare your tick with your buddy's tick . . . see who can grow the largest tick in the shortest time. Eventually the tick will drop off on its own, go away, and lay eggs for the next generation. But not before it has shared nearly everything it acquired from the blood of earlier hosts with you. Ticks do not carry any beneficial partners. You will not be a better person for having shared your life with a tick for a few days or weeks. Get rid of it.

Among the reasons for losing your visiting tick as soon as possible (a list not as long as your arm unless you have a short arm, but nonetheless impressive and only partial):

- ✓ Lyme disease
- ✓ Colorado tick fever
- ✓ Rocky Mountain spotted fever
- ✓ Russian spring encephalitis
- ✓ louping ill
- ✓ Negishi encephalitis
- ✓ Omsk hemorrhagic fever
- ✓ Kyasanur forest disease
- ✓ Langat encephalitis
- ✓ royal farm virus
- ✓ Powassan encephalitis

- ✓ Bhanja virus
- ✓ Nairobi sheep disease
- ✓ relapsing fever
- ✓ tularemia
- ✓ Siberian tick typhus
- ✓ Boutonneuse fever
- ✓ Japanese river fever
- ✓ Tsutsugamushi disease
- ✓ Kemorovo tick fever

The diseases listed above vary from not too serious to invariably fatal. You will have noticed that many on the list take their name from a locality or region: Lyme, Connecticut; Omsk in Siberia; Tulare County, California. One of the characteristics of many tick-carried diseases is that they have, or originally had, a very local distribution. Remember, ticks don't travel far without a host to carry them, and most hosts don't travel much either. As humans travel more, Omsk fever can easily become Anywhere fever. Since a tick can wait for many years for its next meal, the disease it may be carrying might disappear for a long time—but once it gets into a local group of ticks and their hosts, chances are it will eventually find its way to people. Many of the diseases listed are now widely distributed. Rocky Mountain spotted fever is most prevalent in the eastern U.S., due not to the movement of the ticks, but to our own peripatetic habits.

A local disease carried to some distant part of the world can be very serious there, even though it causes only mild discomfort at home. People exposed to a disease often develop some degree of immunity even if they never show disease symptoms. Measles killed thousands of non-Europeans as Europeans carried it around the world. One of the unplanned products of a smaller world is an increased number of diseases that will be serious in a new place. Beware the disease that is named for some place far from where you get it.

If you stay indoors and never leave pavement and sidewalks, chances are you will never encounter a tick and never suffer the potentially damaging effects of Tsutsugamushi disease; but for most of us, that's a very high price to pay for certainty. Better you should be able to recognize a tick and know something about how and when to get rid of it. The only way to eliminate ticks for certain would be to get rid of the outdoors, which makes throwing out the baby with the bath water seem sensible by comparison—and is much too great a cost for what is to nearly everyone a minor inconvenience that can be handled with some knowledge and a pair of fine-pointed tweezers.

Flies *(Diptera)*

The world harbors at least 100,000 different kinds of *Diptera*, including flies, mosquitoes, midges, gnats, and dozens of other groups. What they all have in common is a single pair of wings and unusually large eyes (the better to see you with, my dear). No matter what it seems like, not all of them are in your backyard on the day of the neighborhood picnic. Probably most of the different kinds have stopped off on some human surface at some time. Only a few kinds are of serious interest to humans, and those are mostly the ones for which we have provided food and housing. We have domesticated the housefly, the stable fly, and many other kinds just as surely as we have domesticated dogs or goats, by providing them with an abundance of food and comfortable living quarters (*mi casa, su casa*). Any kind of fly that has acquired a non-Latin name has been close enough to us for long enough to be of interest.

Like many insects, flies of all sorts hatch from eggs and spend some time as immature (meaning before the creature is sexually capable) larvae called maggots. Maggots and adult flies are so different that there is virtually no way to connect the young with the grown-up on the basis of appearance or habits, and for hundreds of years no one did. Both were believed to arise spontaneously from death or decomposition. The larvae are wormlike and certainly appear to develop from nothing but dead stuff. When people first began naming things in the world, all fly maggots were called worms. They still look just like worms, but we have watched carefully enough to connect those unpromising worms with the iridescent flying creatures that they become, and to thus connect the iridescent flying creatures with the eggs that become the worms.

Maggots are terrific feeding machines, but they don't have much capacity to get around in the world. For quite a number of different kinds of flies, the maggots do all the eating while the flying adults are concerned solely with sex and with laying eggs in as many different places as possible. Some adults do require a rich source of food before they can lay eggs (rich in blood or protein—the sorts of things we are made of). The horseflies, stable flies, blackflies, and such that bother us so much are only trying to get together the resources needed to start a family—a noble goal.

The maggots usually feed on something dead and high in protein: something like garbage, dung, dead animals, or, less commonly, live animals. The reason the adult flies are a problem to us and other live animals rests mostly in the limited mental capacity of the maggots. To the maggot, something that is dead on the outside probably seems a lot like something that is dead all the way through. The surface we and other mammals present to the world, that vast plain on which small visitors may take a brief rest, is totally dead. There are no live cells on the surface of us if our eyes and mouth are closed. We are covered totally by a layer, quite thick from the point of view of a fly, of dead, flat, protein-rich cells, penetrated here and there by hair follicles that, like caves, lead directly to the live stuff below the surface. To something with the nervous system of a maggot, we are probably not recognizably different from anything else that is dead and high in protein. They can burrow in, feed, and grow equally well on both us and dead organic matter. Those that prefer us are only slightly specialized.

The adult female fly has a better nervous system, at least more complex, but is under considerable pressure to get those eggs laid. There may be anywhere from hundreds to thousands of eggs developing rapidly in the female and they have to be gotten rid of *somewhere*, and soon. Any woman in the last trimester of pregnancy will sympathize. If the eggs are laid somewhere, anywhere that the maggots can develop, feed, and grow, that constitutes success from the fly's point of view. That somewhere may be dung, garbage, a dead animal, or a live you; they are not all that different on the surface. The specific attraction may be to someplace that is warm and damp—it could be a wound or any of those naturally occurring warm and damp places we all know about even if we don't think of them often. Thinking like a fly might lead you to not letting flies spend any amount of time on your surface, especially near any of those warm, damp places. You don't know which of the flies might be a female (they don't look much different except to another fly) desperately looking for a place to lay a few microscopic eggs.

Most kinds of maggots don't have the eating machinery that would allow them to get under your skin, but out of 100,000 kinds, quite a few do. And just under that dead skin surface is all the high protein food, warmth, and moisture that any maggot could desire. In the more tropical parts of the world, where both life and death are more abundant and varied, there are hundreds of kinds of flies that can develop on human surfaces or just below them. You would of course know you had been visited if you knew what to look for. When you develop significant new lumps on or around some warm, moist part of your anatomy, you should definitely seek medical assistance. Like the ticks, the maggots will eventually (when they have grown to full size) emerge and drop off, but you may not want to wait. Like any young creature with abundant food, maggots grow exponentially, doubling in size and feeding rate in a short time. A maggot barely visible one day can be a half-inch long a few days later. When something is excavating living quarters in you, early removal is desirable and maybe even essential.

One remarkable fly from South America deserves more attention if only for the unique way in which it gets its eggs delivered to its host. *Dermatobia hominis* (anything with hominis as part of its name is trouble for people) does not lay its eggs directly on people, but it gets them there. It overpowers some other, smaller insect (mosquito, smaller fly, any other potential visitor to the human surface) and cements an about-to-hatch egg to its underside. Later, when the unwilling carrier stops for its meal on a human, the egg hatches almost instantly and the maggot needs to crawl only a short distance to a ready-made small wound, where it can take up housekeeping. The advantage of this bullying approach for *Dermatobia* seems to be that it gets its eggs dispersed by energy other than its own and gets them delivered by something less obtrusive.

With their usually limited nervous system, most maggots don't hunt much for food, but there are exceptions. The most notable exception and one of the few maggots that has a definite dependable parasite association with people is the Congo floor maggot, *Auchmeromyia luteola*, which gets its food from sleeping people. It cannot climb up the legs of a bed, but if you are sleeping on an earthen floor in the right part of the world (Central Africa) it can move around enough to find you and suck your blood as you sleep. They seem to be totally dependent on people as a source of food, something not true of any other fly. The Latin name probably comes from what people say when they wake up and realize they have been both host and food for a visitor.

All of us learned early in life that flies carry disease. You don't let them walk on your food because you don't know where they have been before. They could have been crawling on anything, and we all know how germ-ridden "anything" is. Even a fly from the better part of town carries literally millions of bacteria: two million by actual count, done by people who have entirely too much time on their hands. The good news is that none of those bacteria are likely to cause disease for any of us. Most of the germs that cause us trouble live inside of people, where the flies can't get to them. We all have pretty much the same sort of bacteria living on us anyway, and the flies are just carrying them from one person to the next. Beware the fly laying eggs and don't sleep on a dirt floor in Zaire, but cast aside your fear of houseflies bringing disease into your life. Worry about things of greater consequence, like what happened to the stock market yesterday, or whether there is any way to bring peace to the Middle East.

Getting rid of all the flies, viewed as desirable by some of us, would produce more serious problems than any trouble they might cause while here. The world needs flies. Just as a clock won't work well with even a small gear missing, the biological world needs all its parts. The cost of a biologically rich and functional world includes the potentially tiny risk that we might occasionally feed a fly. Mostly they eat dead things and do a remarkable job of keeping the world clean at way less than union scale.

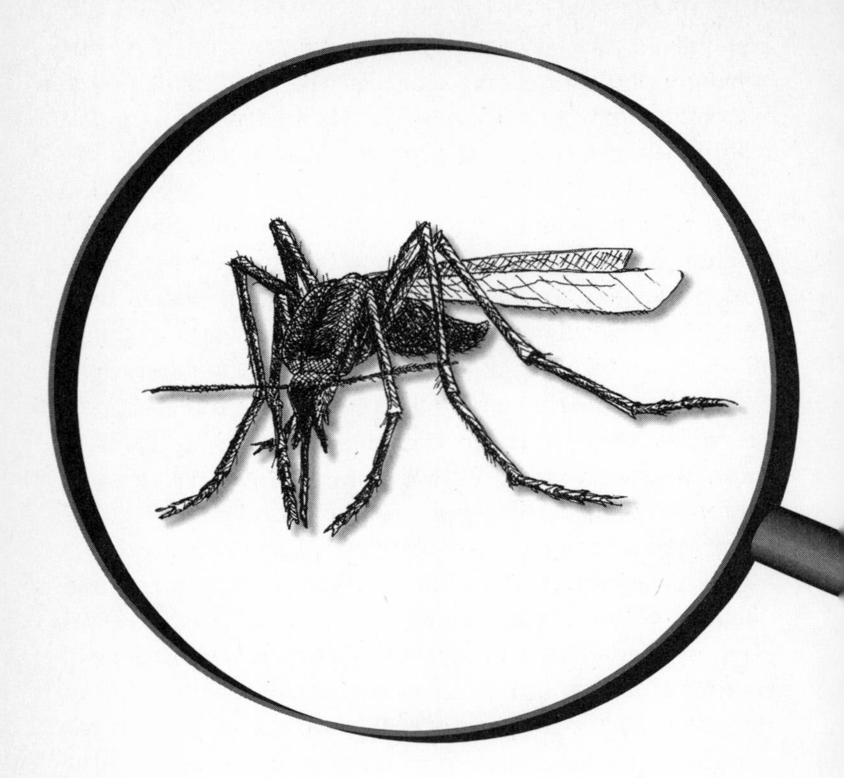

Mosquitoes (Many species)

You are lying in your room. It is a warm summer night. There is one mosquito in the room. You can't see her, but you can hear her. You know it's a "her" because male mosquitoes don't hum like that, and you know that only female mosquitoes bite. You may think because it's dark the mosquito won't find you. Wrong. Mosquitoes love the dark. They have been finding warm bodies in the dark for millions of years. A million years ago the earth already had a nearly perfect, miniature, heat-seeking missile armed with some of the most potent biological weapons known, a missile capable of waiting until its target is most accessible and then striking without a noticeable explosion. The humming stops. The mosquito walks around briefly on your skin, looking for a place where blood is near the surface. You can't feel her tiny feet, but she can feel where your capillaries are and can find them with one of the world's smallest hypodermic needles. If you are asleep you might not even feel her bite. The kinds of mosquitoes best adapted to getting their meals from us seldom touch any nerves, and get away with enough blood to support the development and laying of a sufficient number of eggs to make another hundred mosquitoes. Don't blame the mosquito for trying to provide for her potential family. She needs blood in order to lay eggs, and we have blood—more than enough.

If visiting mosquitoes only carried blood away with them, no one would mind much if they took a little, but just before pumping out the blood, they inject a bit of saliva, and with that saliva can come the tiny disease producers that cause malaria, yellow fever, and a host of less well-known diseases. Only a very few mosquitoes that have already had a meal from a sick person carry these diseases. Mostly we harbor the diseases and the mosquito transports them from a sick person to a well person. In that particular case, it might be considered a good idea to kill the messenger, roughly equivalent to cutting the telegraph lines in time of war to keep the messages of the enemy from getting through.

But second thoughts are always in order. What good are mosquitoes anyway? They are food for other living things (birds, bats, spiders, and other insects), like we all are mostly, or at least eventually. We are food for the mosquitoes and they are food for other creatures we don't even know—a nicely balanced arrangement and one we should generally support. Making sure there are very few people with yellow fever or malaria and that mosquitoes don't get to them is a very good idea. Sleeping with a mosquito net around you is a far better idea than planning to kill all the mosquitoes, and keeping mosquitoes away from anyone with malaria or yellow fever is an even better one. Some parts of the world need mosquitoes, even if we don't personally know what parts. The world would not be a poorer place without yellow fever or malaria, but it would surely be a poorer place without mosquitoes.

If you have no desire to be part of the mosquito's life, is there anything you might do to keep them at bay without destroying all of them? A diet high in garlic will render you undesirable from a mosquito's point of view. What it does to the rest of your social life may be another matter. Researchers are working on ways of extracting the mosquito repellent from garlic and providing it in pill form. We might eventually be able to swallow our mosquito repellent rather than spraying it on the outside.

Chiggers (Many, many species)

Chiggers are almost too small to see, and that's the problem. Chiggers are baby mites, and if an adult is called a mite, you can imagine how small the young ones must be. Even if your eyes are very good indeed, something small enough to hide under a hair is unlikely to attract your attention. It is the activity of chiggers that attracts attention, not their size. It is not an altogether pleasant thought, but chiggers could be swarming all over you in large numbers, and you wouldn't even know about it until they settled down somewhere and began using your skin for food.

Chigger eggs are left lying around on the ground by their mity parents, and the hatchlings are on their own from the moment they enter the world. Having no guidance toward civilized behavior, they take food where they find it and cause no end of trouble. They are the juvenile delinquents of the ectoparasite world. If they survive to become adults, they give up feeding on and annoying other creatures and take to a vegetarian diet. Should an adult find itself on our surface, it lacks both the equipment and the desire to feed. Besides, adults are big enough for us to see and we would likely find them and do away with them.

Since young mites cannot be easily seen without magnification and most of us don't look ourselves over with a microscope, their habits have remained a mystery to most people. Research performed by that very small number of people who do look at their own and others' skin with a microscope has dispelled a lot of the common myths about chigger behavior. The two major questions seem to be: how do chiggers find us, and what do they do once they have?

If chiggers were really looking just for us, their lives would be much more trouble for them and there would be very few chiggers. What they need to get beyond adolescence is a high-protein meal, and anything big enough to cause moving air currents when it passes by is perceived as a potential high-protein meal. Chiggers, like other adolescents without supervision, seem to congregate on what passes for street corners in chigger land. Several dozen could be together on the shaded underside of a grass leaf, where they can hang out for a long while, waiting for a current of air or a whisper of carbon dioxide, both of which announce the passing of a creature big enough to have the resources they need. They will climb aboard anything mammalian (mice, shrews, rabbits, people) or even frogs and lizards if nothing with hair shows up. On people, they probably climb into our clothes first. Once activated, they are remarkably mobile and move rapidly over all the surfaces at hand. Even if chigger-ridden clothes are removed, the chiggers keep moving and looking, and can survive in a pile of unwashed clothes for days. They will continue trying to find the real you if you put the clothes back on again, or even if you put on clothes that were close to the chiggered ones. Chiggers find us by being patient and leaping at opportunities—virtues in many creatures.

What chiggers do when they get to our skin we perceive as much less virtuous. They are most likely to find our skin in those places where skin and clothes are close together: under belts or waistbands or inside socks. For something as small as a baby mite, those places are not confining, just a little quieter and a little damper than the rest of our surface. Once on the skin in just the right place, they spit up a small amount of fluid with enzymes that digest protein (not unlike the stuff you buy for your laundry that magically removes blood and grass stains). Since your surface is mostly protein, the enzymes produce a small pool of already-digested food for the chigger in a few minutes. The next day or so is an endless round of hang on, spit up, and suck up from a gradually deepening pool of food.

Chiggers never actually get under your skin. That idea probably came from some confusion of chigger with the tropical jigger or chigoe, an extremely unkind sort of flea (see page 59). If the chigger only ate from the surface, we would never know it had been there, but it makes something much like a shallow well in our skin, and after a time it gets to the live cells underneath our dead exterior. The live cells respond to what they perceive as an invasion in the same way they would to any small wound; they get irritated. They try to wall off the invasion and at least partly succeed. The result is a structure called a stylostome (stalk-mouth), kind of like the brick walls lining a shallow well. By now the chigger is standing on its head, spitting into the well, and sucking out the dissolved cells from the bottom. The chigger has fattened some with all that good food and is big enough to be noticeable in the center of a red, itching spot. At this stage, most chiggers probably die from our scratching. In any case, they will leave shortly, having acquired enough food to make it to adulthood. The vast majority of chiggers, however, live fast, die young, and leave a really ugly corpse. Only a handful need survive to maturity, because a single chigger can lay hundreds of eggs and a married chigger even more.

Even after the chigger has gone, either to adulthood or to that chigger heaven where no one scratches, the stylostome that we and it have cooperatively constructed persists for days and continues to cause the irritation we usually associate with chiggers. Our *response* to the chigger, rather than the chigger, is the major problem. We make the stylostome, and it itches and irritates. The chigger was just looking for a good meal.

Since there are thousands of slightly different kinds of chiggers, and they are found almost everywhere, is there any way to avoid being food for these young ones? Stay indoors and away from the clothes worn by anyone who has ventured out, or go out only in the winter when chiggers are not around. A better and more practical solution is to plan on a few chigger bites in the course of normal living, but try to avoid the crowds of chiggers found in dry, open, untouched grassy places. And especially, don't lie down in what may be chigger territory. Don't sit down either, unless you have covered appropriate parts of yourself with one of the repellents. Powdered sulfur used to be the material of choice, but DEET (you don't even want to know how the whole word is spelled) works as well and doesn't smell quite as bad. Neither of them kills chiggers (they're repellents), in case you have developed any tender feelings for these misguided youths. Finally, try to think of chiggers as the best hope of the future for the mites that were their parents. It won't help the itching, but you might feel better about living in a world where we are food for other creatures, just as surely as many of them are food for us.

Chapter 3

NEIGHBORS

That group of creatures most closely associated with us historically. They live close by and stop to borrow a bit of food when they need it.

Long ago, when humans wandered around looking for food each day and slept in a different place each night, those ectoparasites that are now our neighbors had to make do with other sources of food and shelter. Present-day animals with those same wandering habits are not usually hosts to fleas or bedbugs. No bed, no bedbugs. As soon as we began to call some place home and return to it on a regular basis, there were our neighbors waiting for us. It was dangerous for most of them to try living right on any creature as curious as the human. We and all other primates have always spent a great deal of time involved in what ethologists call grooming behavior, which means combing through each other's hair and scanning each other's skin looking for anything of interest. You have to suspect that at least some of that sort of behavior had motives other than simple curiosity or the comfort of touch. There might even have been some competition to see who could help groom that cute young thing in the next cave, or that hunk from over the hill. Because of the very real dangers of living right on the things they were living off of, some of our parasites wound up as neighbors, feeding and exploring on the human landscape, but living and breeding someplace else. While an occasional visit to the source of food would be all right for a vigorous adult with outstanding escape mechanisms, the surface of a scratching, looking, and picking human would certainly be no place for those quiet moments alone that all creatures need. As for bringing up a family on the human surface, better the edge of a freeway. It would be safer.

Bedbugs, fleas, and some sorts of lice are only marginally ectoparasites. They are really neighbors for whom we are a convenient source of food. To them we seem always willing to lend a cup of sugar if they are short, and most of them are. The fleas and bedbugs spend time right on us only when we are asleep or otherwise inattentive, grabbing a quick snack and then hopping or crawling back to spouse and family. Most of their more interesting life is spent in the nooks and crannies around us but not on us. The more nooks and crannies, the more tiny neighbors we are likely to have. The more people who sleep near those nooks and crannies, the more varied our neighborhood population is likely to be. Bedbugs are commonly called wall-lice (among other, less complimentary names), because they spend their days out of sight in the cracks of walls or the small holes of a wooden bed frame. Fleas, too, are able to get along without our close companionship for extended periods. These minute neighbors have to be very good indeed at finding us in the dark and at a distance. They necessarily exhibit what might almost be called intelligence in their day-to-day lives. Few would claim to like them, but the volume of literature dealing with our relationship to fleas and bedbugs suggests at least a grudging admiration for their capacity to survive in the face of our most diligent efforts to convert them to endangered species.

Bedbugs *(Cimex lectularius)*

The bedbug is a flattened (to fit in narrow cracks), nearly round insect (six legs) with a sharp beak below its smallest head. It is usually reddish brown and about the size of an uppercase O.

Historically, the bedbug, like us, once lived in a few caves in the Middle East. There it managed to get the blood it needed for food from the resident bats and perhaps from a few cave-dwelling birds. It must have waited on the cave wall or in a small crevice through the night, able to feed only when the bats returned from their nocturnal wanderings and hung it up for the day. Somewhere late in the history of this cozy relationship, humans or something quite a bit like them moved into the caves and brought with them habits different from the resident bats. These new cave dwellers slept at night and wandered out of the caves during the daylight hours. For the period when the cave was shared, and before the nearly human folks moved on to better quarters, the bedbugs (inappropriately named for the time, since their hosts probably used a pile of skins on the floor) must have had a wonderful and abundant time. Sort of a golden age for bedbugs: feed on the bat during the day when it was suspended from the cave wall or roof and was asleep, then, when the bats left at dusk, look around for the first sleeping human and climb aboard. Stay there through the night, feeding and frolicking, and when the bats returned at dawn and the humans began to stir, find a drowsy bat. Next day the same busy story. When the bats were finally driven out by some of the more fastidious housekeepers, the relationship of bedbugs and humans became even closer and the bedbug turned into the cautious creature we see today, spending its days hiding in whatever cracks and seams are available in piles of clothes, walls, rugs, or anywhere else near sleeping quarters, and creeping out at night to feed briefly on whichever human is near. They don't eat much and they leave a very small and only slightly irritating wound.

From the bedbug's point of view, we are just a convenient source of food, close at hand and moving very little for the time they are interested in us. They have not attempted to develop a closer relationship with us. They lay eggs in the debris that surrounds our sleeping places, yet don't use us as a home. In return for their relatively kind treatment of us, we have moved the bedbug around the world, and it is now found nearly anywhere humans sleep. It is, of course, more likely to be found in places where large numbers of humans sleep, either at the same time or sequentially. Until DDT and other effective residual insecticides were available, it was the expected resident of almost any run-down hotel or mission dormitory. While we have by now nearly eliminated the bedbug, we can be confident that it will develop resistance to some or all of the common insecticides and that only diligent destruction of its hiding places will keep it at bay. We have been too good to it for too long for it to abandon us now.

How would you know that the bedbugs had moved back? It requires coordination. Go to bed. Lie still until you think they have found you, then simultaneously leap out of bed, throw back the covers and turn on the lights. You should be able to see them before they get back under cover. If this seems too much for you, just go to sleep, and when you wake, look for feeding wounds. The bite of a bedbug appears not very different from that of any other blood sucking insect: a raised red spot centered on a very small puncture. The bites will most likely be on the legs and are often in rows of three or four: sometimes it's hard for a feeding bedbug to find just the right spot. The better signs of the presence of bedbugs come from the places where they spend most of their time, that is, the tiny nooks and crannies of walls and furniture. They leave dark spots on walls and beds, spots that are smaller and more long lasting that those they leave on people. After a meal, lacking civilized comforts, they defecate wherever they find themselves. And they smell—an organic smell, partly from the bugs and partly from the unclean conditions that make it possible for them to live and raise their little ones. The young bedbugs don't use us for food, but find what they need in the small amounts of dirt and refuse around our sleeping places. This dirt and refuse includes pizza, old potato chips, and disgustingly, the partially digested blood defecated by the parents of the baby bugs. I presume that bedbugs have not yet found my children's rooms, since the walls, while not spotless, lack bedbug spots. They would certainly have felt right at home had they arrived. The rooms of the majority of adolescents would have more than enough nooks, crannies, and old food. In most of our modern, clean, and painted rooms bedbugs are blessedly rare. They do, however, have that enormous capacity to reproduce that all bugs share, and with abundant food for both babies (old pizza) and adults (us), they can reestablish themselves quickly.

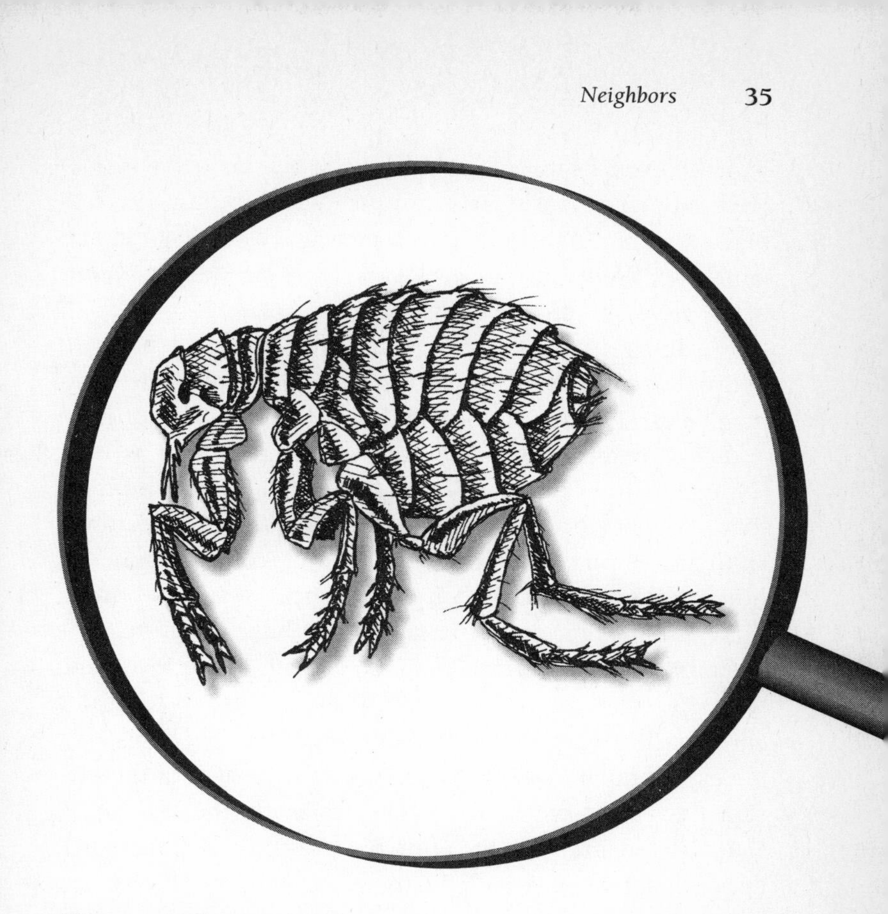

Fleas *(Pulex irritans)*

If we were anything approximating the size of a flea (slightly smaller than a bedbug), we would without question consider them superheroes. They would seemingly be able to disappear with a great leap from wherever they were and reappear hundreds of yards away. They'd be able to jump high enough for us to wonder if they had left the earth altogether. If we could teach them to handle a basketball, the NBA would never be the same. And not just superstrong, but supersensual as well (a trait shared by few superheroes, who may be able to leap tall buildings but seldom jump anything else). They bear what is generally acknowledged to be the most elaborate sexual apparatus known among the animals and they copulate for as long as four or five hours at a stretch. Even from our lordly perspective as people, fleas have been given a nearly unique place among our tiny neighbors. Not many of our ectoparasites have major streets named for them, such as the Alameda de las Pulgas (Avenue of the Fleas), south of San Francisco.

Judging from the things people have said in print about fleas over the centuries that they have been our neighbors, they are, if not godlike, at least among our most enduring companions. They have been described as jolly, impudent, fun loving, avid, and other adjectives implying that the behavior of fleas is at least suggestive, if not downright lascivious. That they live exclusively on a diet of human blood when they are adults seems less important than our perception of them. Lovers have suggested the desirability of wandering over the surface of the loved and taking a small nip where they wish, in exactly the same way that fleas do; and stories of "What the Flea Saw" were once as popular as the heaving bosom romances of the present day. What we have written has made the flea bigger than life in many ways (not too difficult), although it appears not to have helped the well-being of the human flea, which is practically gone from the world's cleaner households.

The Latin *Pulex* comes from a word that means dust, and back when people regularly considered that vermin of all sorts rose spontaneously from the earth, the flea as animated dust made more sense than other explanations. While it was true that the typical flea was larger than most dust, and its animation far greater, the place one was more likely to notice fleas was in the dust of a room or space that had not been occupied for some time. Leeuwenhoek, a seventeenth-century Dutch microscopist, was the first to follow fleas from egg to juvenile, to adult and back to egg, by keeping them on his body and watching closely with his homemade microscopes. He dispelled the dust-to-flea notion. Present-day flea researchers still feed their own fleas in little cages strapped to their ankles. What hatches from the flea egg looks not at all flealike and does not hop or even move with any facility at all. It lives for a few days or weeks as a small, wormlike creature, feeding on dirt, dander (see page 48), debris, and its parents' defecations until it spins itself a sort of cocoon and transforms itself into what we think of as a flea. The adult, shaped like a sunfish with six legs, is flat enough to slip easily between hairs in even the densest thicket, and hard enough to resist being crushed between the fingers of the most determined host. You have to bite them to kill them most of the time, which explains a lot of what the monkeys at the zoo are doing with the tiny things they find on each other. The adult flea needs food as soon as it emerges from the cocoon, and unless it finds itself in the middle of a crowded housing development, starvation is a grim possibility. Many types of fleas are able to wait for a year or more in the cocoon, and will emerge only when there is some vibration, heat, smell, or other indication that a person has come by. To the person entering a long-abandoned former home of flea-keepers, it might well seem that the dust itself was coming to life. The adult emerges quickly and brings to bear its considerable food-finding capacity, seeming to rise up from the dust of a long-closed room.

Exactly how such tiny ones find and leap aboard some blood-bearing creature varies from one kind of flea to another. Some fleas can jump one hundred times their own length, and are capable of repeating such jumps hundreds of times per hour for days. Rabbit fleas, which have been studied extensively, are able to find rabbits very well indeed. Nearly half of 270 marked fleas released into a large fenced field with three rabbits were found on the rabbits in a few days; a nearly supernatural achievement considering the size of the field, the size of the fleas, the mobility of the rabbits, and the fact that fleas see poorly over anything but very short distances.

The human flea feeds on us, pigs, and badgers with comfort and success, although it will, if starved, take an occasional meal from something else. There is an evolutionary and social lesson to be learned from that peculiar combination of hosts. Creatures that wander as they feed during the day and sleep in different places each night might have no fleas. That includes most of the grazing animals and all of the apes and monkeys. We probably did not feed any fleas until we settled down and began living in the same place for weeks or months. Growing from egg to adult for most fleas takes as little as one or two weeks, but during that time any potential host that moves to a new place daily is long gone. Only the fleas that associated with animals that came back to a burrow or nest had very many successful offspring, and those offspring passed along their parents' habits. Badgers or their close relatives may have been the original hosts for our fleas, because neither pigs nor our primate ancestors returned regularly to a home base, as badgers do. Once our flealess forebears set up permanent housekeeping somewhere and began to sleep in close to the same place most nights, they became party to whatever fleas found us tasty and were agile enough to escape our efforts to eliminate them. When we started to keep pigs we most likely shared quarters with them for a while, just to keep them safe from harm, and they most likely picked up the fleas from us. Fleas are more associated with habitat than they are with particular hosts, and even now, human fleas are flexible enough to temporarily use rats or cats or dogs for food, even though they have a strong preference for us. Rat or cat or dog fleas will return the favor.

At least some sorts of fleas depend on the sexual cycles of their hosts for their own reproduction. Rabbit fleas can only lay fertile eggs when the rabbit they feed on is pregnant. Quite a nice arrangement, since it assures that there will be new hosts for the new fleas, and new fleas for each young rabbit. No such arrangement is known for human fleas, but they are commonly acknowledged to be more attracted to ladies, especially young ladies, causing some to suggest that the blood of young human females must give a special kick to the wandering flea.

How have people over the ages, especially the ages before DDT, coped with the presence and problem of fleas? Any way they could. It has been suggested that those tiny, yappy, otherwise useless lapdogs were at least partly bred as flea traps. "Take Fu-fu away now. Brush him and given him a good bath," while not as effective as insecticide, was certainly somewhat successful in keeping My Lady's flea population at a minimum. Especially since My Lady probably bathed rarely or not at all. It is only a short step from a lapdog to a fur piece around My Lady's neck, and if not worn constantly, it would have been useful in misdirecting fleas. Our little scurrying neighbors would find it a great place to hide, and even if there was no food available from it, they might hang out long enough for the fur piece to be back in the closet. There were likely only a few fleas on any one person at any one time, and getting rid of even one or two of them would have reduced scratching. A fleabite, while not really painful, is at least disturbing, and is likely to itch for a few days.

Where are we now in the ages-long saga of fleas and people? The flea circuses that so entertained generations of young and old have closed down because it has become almost impossible to find enough human fleas, and other kinds of fleas just don't perform well. Finding amenable transportation, housing, and food for the performers is also more difficult. Few of us want to be the only one in the neighborhood with fleabites, even if it gets us close to show business. Most of the fleabites reported on humans at present are from cat or dog fleas just trying to accumulate enough food to get them from one pet to the next. The vast majority of the human population would not likely recognize a human flea if one should appear. It is a shame in a way. Everyone should have the opportunity at least once to step into a long-closed, dusty, formerly flea-ridden room and see the dust apparently animate itself and head for their ankles in great leaps. That experience would do more to develop genuine reverence for life than any number of television specials on the strange habits of rain forest birds or bees.

Body Lice *(Pediculus humanus corporis)*

There are more than four hundred different kinds of lice that live on mammals, and nearly every mammal has its personal kind. Lice are choosy about their neighborhood and aren't comfortable moving from one kind of host to another. Even if you have a lousy dog (one that bites people, for example) you don't have to worry about acquiring any of its lice. All kinds of lice but one live right on the thing they are living off of. The exception is our own personal neighbor, the human body louse. It almost certainly used to live right on us, as a couple of its close relatives (the head louse and the pubic louse) still do, but we have changed through our history with the louse, and the lice that didn't abandon us have had to adjust.

Lice come as close to living right on us as they can, but our hairless or much less hairy surfaces are just too warm and bright for their continued comfort and reproduction. On our earliest ancestors, lice probably spent most of their time away from the skin surface, in the hair—kind of like monkeys living in the upper reaches of the forest that covers our skin. The lice on real monkeys or apes are similar to those on us, and we probably inherited them from our earliest ancestors. A few thousand generations ago, we were a lot hairier than we are now. Even in the sixties we were not as hairy as we used to be. During those earlier, simpler times, the body louse lived right on us, held on tightly to our hairs, reveled in the shade cast by that abundant forest, laid its eggs on the hairs, and when it needed food (several times a day) it clambered down to the skin surface, found a sheltered spot, and had a good meal. Separated from its meal ticket, a body louse has about four days to starvation. The louse was well-adjusted to its specialized life and could change the microclimate it lived in by just moving out farther in the hair during the summer heat and a bit closer to the surface during the cold spells. With a warm surface so close, and the mostly cooler air all around, moving a half-inch out into the hair was the equivalent of moving from the gulf coast to the mountains—winter on the nude beach and summer at the mountain resort.

That idyllic existence changed rapidly when we began to lose the hair that covered a large part of our bodies. First the hair probably got thinner and farther apart. If you are a louse you have to be able to reach at least two hairs to hold on well, and for our body lice the critical spacing is about a millimeter. As those forests of hair became sparser and then declined to the scattered copses we now bear, the lice were forced into confined quarters where they probably never saw the lice from the next forest over. Finally, from the point of view of the louse, there were only scattered places with enough hairs close enough together. It was a terrible time.

There were other, burlier, meaner lice that lived where the hairs were farther apart, and the gentle body louse was forced out of its native home onto the bleak, brightly lit open plains of the evolving human, where it was easy prey to its sharp-eyed host, where there was no place to hide, no sturdy hairs to hang on to, and where it was too hot for successful reproduction, even if you could find a quiet place. A few lice took up permanent resident in one or another of the still-hairy locations, but most were faced with almost total loss of suitable habitat. If the lice had had an Environmental Protection Agency, they might have had a chance, and we would probably have been required to remain hairy.

Fortunately for the louse, at the same time people were losing hair, some of the people were moving into colder climates and were beginning to wear furs—furs from other mammals. Lice could live away from the human surface in our new furry clothing, where it was just cooler enough, where there were plenty of hairs to hang on to, and where the food store was just a short crawl away. The furs were probably never removed until spring and that was at least several louse generations. People may have slept on the furs through the summer and kept at least a few lice happy until fall. When people mostly gave up furs for knitted or woven clothes (mostly made from animal hairs or vegetable fibers of the right diameter and looseness of fabric weave) our lice made other small adjustments and began to live where they live now, in the seams and creases of our clothes. Kind of like a high-rise above the mean streets where the louse had to make a living.

Were we to stop wearing clothes permanently, or even if everyone removed them daily, the louse's last refuge would be gone and so would it. No clothes, no body lice. A simple solution, but one that doesn't work well where it snows. Next-best advice to avoid lice: take the clothes off at least occasionally and wash them in hot water. That every Monday was enshrined as wash day may have had as much to do with how long it takes louse eggs to hatch as it does with a calendar. Unlike fleas, lice cannot move on to another host to tide them over. *Humanus* in the louse's name means what it says. They are faithful, even dependable, and have managed to make the adjustment to polyester. They may be less sophisticated and less romantic than our fleas, but they have been with us longer than we have been people.

Like anything that uses us for food, lice cause some irritation. A textbook entitled *Entomology in Humans and Animal Health* by R. F. Harwood and M. T. Jones (Macmillan, 1979) declares that louse bites may produce "general tiredness, irritability, depression, and pessimism," an understandable set of reactions. When nearly everyone slept in their clothes, didn't wash their clothes or themselves for months or years at a time, and lice were everywhere.The literati—from Chaucer right up into the nineteenth century—trying for something positive, talked about the louse being noted for its affection for man (understandable), its retiring nature (essential), and its faithfulness. They made lice sound like neighbors we would all appreciate, even if they stole a little blood occasionally.

The eggs of lice are commonly called nits and are right at the border of visibility. In the case of the body louse, they are attached to the seams or creases of clothing. They normally hatch in a week, depending on temperature. The louse-to-be hatches itself by taking in air at its head end, and passing it along through the gut, where it accumulates at the closed end of the egg and helps to force the young louse out and into your clothes. The just-hatched louse looks almost exactly like its parents except for size. It is almost too small to see, is nearly transparent, and is thus safe from any but the most diligent examination by its host. The female louse lays many more eggs than do most other ectoparasites, probably because its home in our clothing is a little more precarious than that of other neighbors. Our louse can lay up to three hundred eggs in its month-long life, all of which can hatch in a week and be ready to lay eggs themselves in another week. Within a month, one female might have a few thousand descendants. Some diligence in washing, cleaning, and searching our clothing and bodies would certainly seem to be warranted.

Wouldn't a few thousand lice crawling around and feeding regularly be cause for action? It depends on where your attention is focused. Reverence for life carried to an extreme or just a lack of care for oneself can have strange outcomes. For holy men in some cultures, feeding and housing the poor took the form of maintaining a thriving population of poor, otherwise homeless lice. The more lice, the greater the holiness. People, holy or not, often develop immunity to the bites and irritation of lice and so can house thousands with minimal discomfort. The only long-term effect of such patronage was the development of deeply pigmented skin, a condition called vagabond's disease or vagrant's disease. There may well be a basis in the history of Europeans and our parasites for the not-uncommon negative response to swarthy folks from another country.

What have people done to keep the numbers of our louse neighbors from getting excessive? Practically everything. For most of human history, the most common diseases and conditions had the most numerous (generally useless) cures and treatments, and before DDT and detergents, almost everything was used to control lice (odorous plants like louseworts or fleabanes spread on the floor, similarly odorous plants spread on people, odorous unguents, incense, prayers, and curses). Before Leeuwenhoek and other early microscopists followed lice from egg through growth back to eggs, lice seemed to appear from nowhere, and efforts to make them go away logically included incantations. Once we knew what the eggs were, where they were, and what they became, control was more rational. A weekly wash day probably did a lot to make the life of the louse more chancy. Clothes stored for a month away from people became louse free through starvation of the lice, which may explain the usual nineteenth-century practice of storing out-of-season garments in closed trunks.

What is the future for the body louse? In the absence of social chaos, most people will probably never see or feed a body louse. It's as good a reason to prevent wars and riots as I can think of. Lice develop resistance to most insecticides that are not directly harmful to people, and there is some evidence that lice that are resistant to DDT lay more eggs than those that do not. In the long run, a peaceful world will do more to hold off lice and the diseases they might carry than any amount of insect control exercised after the fact. Epidemics of typhus and other louse-borne fevers are a high price to pay for getting your political agenda addressed.

Dust Mites *(Dermatophagoides farinae* and *Dermatophagoides pteronyssinus)*

Where does all that dust come from? It piles up in all the corners and makes fluffy little dust bunnies under the beds. It hangs in festoons from neglected ceilings. It doesn't blow in from outside because many of the places it accumulates are nearly sealed from the out-of-doors. There is no less of it in the summer, when windows might be open, than there is in the winter, when they are closed. There may even be more of it in the winter. Few clues about its origin make sense until we look at the dust with a microscope. Besides the small and large hairs that both we and our pets shed at a pretty constant rate, much of the dust consists of small flat plates, usually six sided and slightly wrinkled on the surface.

They are skin cells, mostly our skin cells. We shed them constantly, and in unbelievable numbers. They are the body's way of keeping itself clean and free of invaders. If we could see just slightly smaller things, each of us would seem to be walking around in a cloudy halo of shed skin cells. When we say "Pardon my dust," we speak a truth more profound than we know. Dander is the formal name for this cloud, and the only time we are likely to take notice is when we wear something black and someone points out those white flakes on our shoulders. "Dandruff," we say, and seem to imply that it's some sort of diseased condition that the right shampoo will eliminate. Dandruff won't go away. We shed skin cells all the time from both head and body. And not just skin cells, but tiny, nearly invisible hairs from every surface except our palms and the soles of our feet, making a constant contribution to that never-ending rain of dust settling into all the corners and thinly covering every place we sit or stand.

Besides being among the smaller free-living creatures, and among the more numerous on earth, the many kinds of mites are users of practically anything that might be called food. There are cheese mites, flour mites, grain mites, and thousands of other kinds. Then there are dust mites. What do dust mites eat? Dust, our dust, those thousands of skin cells we shed into the world around us. Dust mites live almost everywhere that we do, too small to be seen. They live in our carpets, in our furniture, in and on our mattresses—everywhere. Any place where we shed and the relative humidity is at least sixty percent, dust mites can thrive. Think of how much dust there would be if the mites weren't eating some of it.

Do dust mites sometimes eat our skin cells before they are shed? Do they live right on us? They may have tried at some point in their history, but living right on people is dangerous for anything, even something as small as a mite, and the ones that tried probably didn't live as long and had fewer baby dust mites than their cousins who lived in the safer neighborhoods. Dander was everywhere; no reason to try to wrestle a person for it. Result, *Dermatophagoides* (skin eater) mites who don't live right on our skin, but certainly live off it.

One small problem: it is possible to unknowingly inhale a mite or a mite part (mite parts are more common than whole live mites, since there are no mite graveyards). Or you could get a mite or a mite part rubbed into your skin. Many of us have, as a result of such contacts, developed a sort of allergy to dust mites. If someone tells you that you are allergic to house dust, what they really mean is that you are allergic to dust mites. Wouldn't it be a good idea in the light of this minor inconvenience to just get rid of the little dander eaters? Good luck. There really are no insecticides (remember mites are not insects) that cause much difficulty for the dust mites. Spray as much as you like—they won't be bothered, but you might. Are there miticides? Yes, but not of the sort that you would want to spray around your house without risking doing yourself more damage than the mites. Might as well live with them and try to forget they are there. Or you could consider them small vacuum cleaners that work for almost nothing. Dust to dust, mites to mites, mites to dust, and dust to mites.

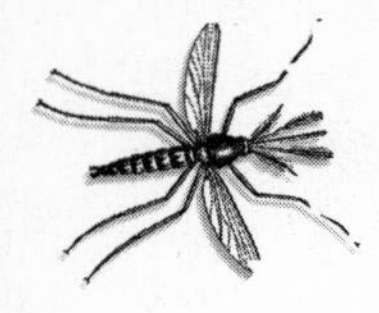

Chapter 4

RESIDENTS

That small number of creatures that live right on us constantly.

With all our activity and curious searching, the number of creatures who take up more or less permanent residence right on our surfaces is pretty small. Some live in the hairy places where they can be out of sight most of the time, and a few burrow in just far enough to keep from getting washed away by either the morning shower or the flood of sweat after the evening jog. As soon as we lost most of the hair that we inherited from our primate ancestors, all that bare skin surface was just too dangerous to live right on. Some of our residents are so small that even the keenest-sighted humans have never been able to see them, but even those could be brushed off or washed off if they didn't have a good place to hang on. They live below the surface for the same reason that a gopher spends most of its time in a burrow and the gambler sits with his back to the wall. The too-small-to-see burrowers that caused some sort of visible skin reaction must have been a source of consternation from the earliest time we were capable of consternation until the time when microscopes were first available and we began to turn them on ourselves. Diseases and conditions that had been mysteries or gifts from the gods were finally seen as having ordinary causes. With the knowledge of these tiny companions came the creepy feeling that we were more than just the sum of our parts; we were at least partly the sum of other creatures' parts.

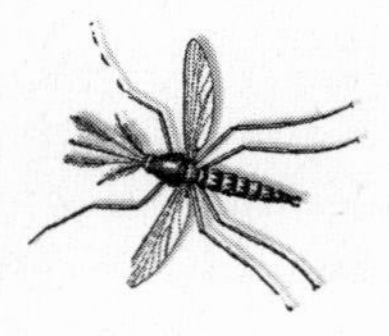

Head Lice *(Pediculus humanus capitis)*

You will acquire and maintain body lice by living in your clothes and washing them less than once a month. You get head lice by going to school. The residential form of our native louse is just as affectionate and faithful as their very close relative, the body lice, but this particular pediculus is pedophilic—its affection is most often directed at children, especially young girls. The usual pattern of closeness and sharing that is characteristic of children seems made to order for the perpetuation of the head louse. Téte-à-téte is not only a good way to learn secrets, it is also a good way to get head lice.

When its environment changes dramatically, any organism must change or die. As the human body lost its covering of long hair, the lice separated and took of different lifestyles. In evolutionary terms this all happened only yesterday, and head and body lice are not yet very different. Those on the head are only slightly smaller and have only slightly different habits. Like nearly all lice, they are somewhat slow and secretive. They need to move less than body lice. They are pale, as befits a creature that avoids the sun at all costs. Like body lice, the young ones are nearly transparent, but are even smaller. Living on the head makes it very difficult for their owner to see them. Friends need to tell you when you have head lice.

The eggs of the head louse are glued to hairs just above the scalp, where they remain even after the enclosed louseling has left. Since head hair grows more or less constantly, the location of the abandoned nits provides some information about just how long the little creatures have been around. The just-hatched young repair to the scalp surface, where their first meal is only a few cells down and readily available. They can hang on to the surrounding hairs with claws and clamps even through a thorough washing, and are likely to be dislodged only with special soaps that contain some amount of hexachloro-something.

The presence of body lice, even when connected with a high degree of holiness, is usually associated with personal habits that are considered slovenly in our cleanliness-conscious culture. Head lice are commonly related to who you spend time with rather than your approach to soap and water. Head lice positively thrive on a clean scalp, and unpleasant as they are, they should not be directly connected to filthy habits. While lice don't commonly move much, they can move enough to get from one person to another if the persons' heads are close together or if they share caps or earmuffs. And the usually longer hair of younger girls is likely the reason for the greater incidence in them, not any warped proclivity on the part of the lice. With more hair there are more places to hide and more to hang on to. The beginning of adolescence seems to have a negative effect on the head louse, one of the few really positive things about the beginning of adolescence.

Even with the enormous capacity of lice to lay eggs and potentially hatch little ones, some diligence in washing and combing probably keeps the numbers low in most people who have them. The risks to the louse get greater as it grows, becomes more visible, and irritates more, so most of the resident population at any one time will be the nearly transparent young that are only a few days out of the egg. All an adult has to do to perpetuate the population is to replace itself, and if it can lay a few hundred eggs, that means that only one percent or less of the total louse load should be expected to live to reproductive age. Life for a louse may be characterized by a warm climate, abundant, comfortable habitat, and instant food availability, but life is obviously chancy.

Head lice have only one predator to worry about—us. We are large, we can see well, and we spend a lot of time looking at most of the parts of us and others that we can see. Life for a head louse is not all darkness and afternoon cocktails. They probably live more like the street waifs in a bombed out city, hiding in the shadows and grabbing a bite when they can. On those grounds we might legitimately feel just a little sorry for the state to which we have reduced the head louse; struggling for existence on the small island or suitable habitat left to it, trying desperately to avoid the attention of that great eye in the sky that appears just before the great hand that plucks them up and out of existence.

Pubic Lice *(Phthirus pubis)*

During the gradual evolutionary separation of suitable hairy louse environments on humans, the pubic louse, also known as the crab louse, took up as its rightful location those places where hairs were just a bit farther apart. The crab louse is about two millimeters wide and is best able to hold onto hairs that are separated by that distance. That limits it mostly to armpits, eyebrows, eyelashes, and a few other places that are harder to see. Its nearest relative lives on gorillas, who presumably have hairs somewhere on them that are separated by about two millimeters. Unlike the relatively fragile, pale-looking lice of the rest of the body or head, the pubic louse is almost burly in appearance, broad shouldered and with obviously muscular arms or legs or whatever all six of them are. They seem designed to be able to hold on more securely than other lice. Why they would need to hold on more securely is worth some thought. Batting eyelashes might be a reason, as might also the rapidly raised eyebrow. It seems most likely that those individuals living in the more extensive pastures farther south, where they are regularly subject to the convulsive earthquakes of passion, probably need all the legs, hooks, and clamps they have to hold their place.

The crab louse benefits from a warmer and wetter climate than its relatives. It prefers high humidity, and easily survives and thrives at temperatures that would render head or body lice inactive or dead. It is also much more of a stay-at-home, probably as a response to the smaller areas where it finds appropriate conditions, and it will in fact starve within hours if removed from its host. Once it has found just the right climate, it generally remains in one spot, moving only to locate a consenting crab louse of the opposite sex. Otherwise it feeds and defecates in that one location until it is found by its host or dies of old age.

Considering its close association with places and activities venereal, the crab louse itself isn't much for sex. Its sluggish nature and strong tendency to stay in one place pretty much precludes recreational participation. Its total reproductive capacity is extremely low. The female produces very few eggs by comparison with practically any other ectoparasite. How it survives so well in the light of our continuing efforts to get rid of it remains something of a mystery. An individual crab louse may move only a half inch or less from where it was hatched to where it more or less permanently attaches itself. I suppose there are fewer risks if a creature stays closer to home and if home is a place that people are not particularly inclined to look. Grooming behavior for all sorts of primates including us extends to most of the body surface, but usually doesn't concentrate much on the nether regions. I suppose our ancestral primates didn't want to be misunderstood any more than we do.

If crab lice stayed totally at home they wouldn't have much of a future. They do move, but just a little way and only when they are disturbed. Yes, it is possible to get them from a toilet seat, but only if you were preceded there by someone with disturbed pubic lice, and only if you stayed there long enough to read ten pages of the *Reader's Digest*. No relationship between chronic constipation or magazine subscriptions and crab lice has been established, but it might be worth looking for one. Their presence in armpits and eyebrows is not the product of their long migration over the hairless deserts between the groin and the promised land. If you have crab lice in your eyebrows, you probably earned them.

The affection sometimes expressed for other lice and fleas seems not to have transferred to *Phthirus pubis*. It has not been saluted in the same ways. No one has suggested that it is affectionate, although its life is closely tied up with affection. No one has suggested that it is faithful, even though it stays closer to host and home than any of its cousins. We just haven't been able to identify with the crab louse. It seems not to suffer from our rejection, and has managed to become the most common and widely distributed of the human lice by far. Its life cycle seems just long enough for us to have forgotten where we might have been when we acquired it, and our habits are such that there has never been a shortage of new territories to conquer. How close can two people get? Close enough for the laziest, least-likely-to-move louse of them all to get from one to the other. If we moved quicker, there might not be any pubic lice. If they moved quicker, there would be way too many. As with Goldilocks and the bed, the relationship of *Phtirus pubis* to *Homo sapiens* seems to be just right, at least for the louse.

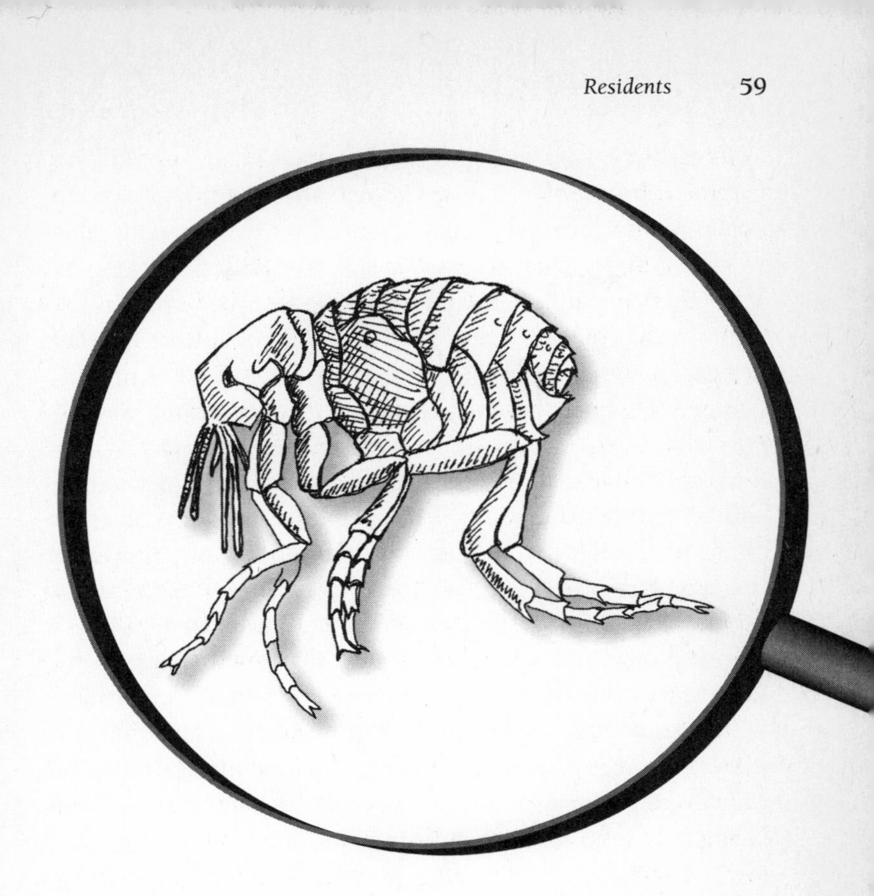

Tropical Chigoe or Jigger

(Tunga penetrans)

It may well be that the reason many people suffering from or just talking about temperate zone chiggers (larval mites, see p. 25) believe that those little ones are able to burrow under their skin is the partly forgotten and partly misunderstood connection to a tropical flea that does burrow right in. Just a few letters can make a remarkable difference. Chiggers you can live with; chigoes you don't want to have anywhere too close. If you live in climatically temperate places in the world, don't worry. If you have to travel to or live in tropical America, keep your shoes on whenever you can, and learn to watch for the signs that *Tunga penetrans* (that last name says it all) has moved in.

Size, especially small size, is important to the survival and success of many of our ectoparasites. The chigoe is at least in contention for being the smallest and thus the least noticeable of all the fleas. The usual adult size is a millimeter long or less—that's just a little more than the size of the period at the end of this sentence. Combine that small size with the usual flea capabilities of jumping and finding hosts, and it is not difficult to see how the chigoe manages to get along. Like all fleas, the chigoe hatches from an egg and spends some time as a wormlike larva, living on dead organic matter and debris and being no threat to any other creature. Once it becomes an adult, with adult capabilities and adult appetites, it feeds on the blood of birds and mammals. Even this is not really a problem because it doesn't eat much, and being small it doesn't cause much discomfort. The problems begin after copulation, a not uncommon circumstance for many creatures. Following copulation, the male dies, its purpose fulfilled. The female becomes extremely active in searching for a host (probably looking for new and unusual kinds of food). Making and laying eggs requires a lot of energy, and that means that a lot of food is necessary. Bringing all the usual flea-like leaping and host-finding capacities to bear, the female vigorously explores its usually sandy home until it locates and leaps aboard some suitable passing animal. It could be a chicken, a pig, a mouse, a rat, or the creature that provides homes for all those other hosts, us. Once on a potential host, the female does not explore much, but settles into the first crease or crack small enough to hide her. There she burrows head first into the surface and feeds continuously. Provided with an endless supply of food and an out-of-the-way corner, usually between the toes, around the toenails, or in a crack in the bottom of the foot, the female grows rapidly and begins to lay eggs. The eggs are simply dropped off into the surrounding environment in large numbers as the host walks around. The host's skin tries to reject the chigoe, much as it would try to reject a splinter or small thorn, but the female continues to burrow and grow. She may reach the size of a small pea, and will lay thousands of eggs. She is eventually rejected by the host, but not before she has had the chance to travel extensively with her host, and the chance to deposit offspring in thousands of places, some of which are likely to be just right for the young to develop. Places in farmyards are ideal, with abundant potential hosts and abundant organic material.

If the female grew a little less, we would probably not be aware of her, but the combination of her size, her feeding, and our attempts at rejection result in a serious wound and considerable chance of infection. Harboring a pea-sized parasite under the edge of a toenail is most often only a minor, if painful, inconvenience for us, although it could in extreme cases result in the loss of a toe. For something the size of a mouse, the chigoe is probably life threatening.

Chigoes live only in South or Central America, and have shown no tendency to move farther north. We are not essential to them, but are a convenient host if we happen to be in the right place at the right time. Should you travel in parts of the world where chigoes might be, keep your shoes on and don't sit down in places where there is sandy soil, chickens, pigs, rats, or mice. Keep a careful eye on all the creases and crevices on your feet. Remove the female at the first sign of an irritated spot on or between your toes. The irritation means your body is trying to reject her. Listen to your body and help it in this noble work. Unless you are extraordinarily curious, you should probably not wait to see how things develop. Since the chigoe flea can get along without us, we should encourage that route for the anxious pregnant female. It is not an abandonment of social responsibility to suggest that she take up residence on some other creature's foot.

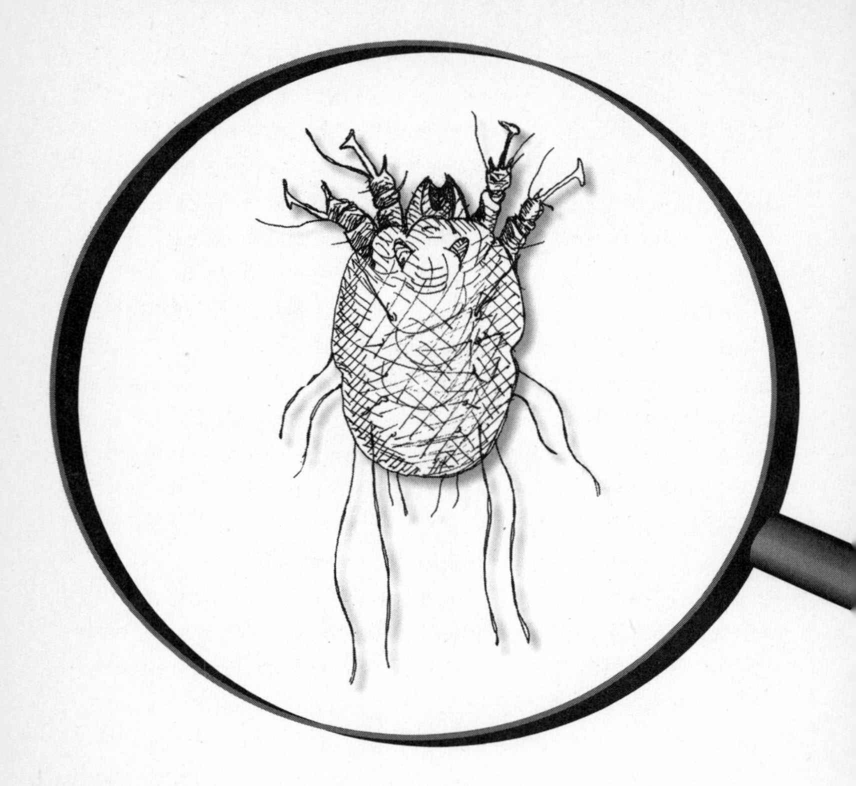

Itch Mites *(Sarcoptes scabiei)*

Before microscopes, mange, or scabies, was a disease without apparent cause. It seemed to occur most often in poor and poorly nourished people. In Sweden it was called Norwegian scabies, a term that persisted well into the twentieth century. In much of Europe outside of Poland it was called *Plica Polonica*. People in Wisconsin called it the Michigan itch, and to those outside the military it became the army itch. It has also been called the seven-year itch in honor of nothing but its capacity to persist. What Norway, Poland, Michigan, and the military have in common is people sleeping in crowded quarters and thus sharing one of our smallest residents. Scabies mites probably infest hundreds of millions of people throughout the world, with no relationship to ethnic origin or location. Only poverty seems associated with the itch, and that only because of the closeness of sleeping places in conditions of limited housing.

Mange mites are too small to see; small enough to live right on and even in our surface. Even the largest are not a half a millimeter in length. They are not pigmented, and many of them spend most of their time just below the surface. Before we could see them with microscopes, all we knew was that there was an itch on our elbows, wrists, knees or some other dry spot with skin folds and creases, and that the itch was worst at night. Now that we can see them we know that the mites are roughly oval, with the eight legs that all mites have, and with a very efficient set of jaws in front that make it possible for them to burrow in and completely out of sight in a matter of minutes. They are pretty much the body's equivalent of a mole in your personal lawn. Like the moles, they are very well adapted to living in their hazardous habitat, and in spite of our best efforts to eliminate them, they persist.

Only the mature females burrow, and they only burrow into the layer of dead skin on the surface. The itch seems to be the result of the mite's lack of personal hygiene (it defecates where it lives). In their burrow, females go about the female business of laying eggs that hatch in a few days into young mites that look like miniature editions of their parents. The young mites live in the burrow for a few days and later wander around the surface. As adolescents, they are still small enough to rest in a hair follicle, and like other adolescents, they eat whatever they find in their environment: dead skin cells, fat-rich oils, and secretions. As we have seen, our skin is a land of milk and honey if you are small enough to escape notice. With maturity, the thoughts of females seem to turn to settling down. As is true with many creatures of all sizes, the males continue their wandering ways for their whole life. Males live on a surface that is regularly washed, flooded with sweat, dried, rubbed, pressed, and folded. Mature males are rare. Fortunately for the mites, only a few seem to be needed. Copulation of mange mites has almost never been observed, although it must occur with some frequency. Either they are very discreet or we are not very observant. The survival rate for the young is also low, since while on the skin surface they are even more subject to the floods, droughts, and earthquakes of their uncertain homeland. Their front feet have suckers that help in holding on, but the scratching and itching they cause produce minor swelling and enough additional fluid buildup to wash them away or drown them. Only in the almost total absence of scratching, washing, or sweating does the number of mites on any one person exceed a few dozen. Most of the reason we pay attention at all to the mange mite is the itch they produce on most people. If they were a little better adapted to us and caused less itching, we would never know they were there.

The female mites die in the one-way burrows they have made. They appear not to be able to back out and, anyway, there are all those young ones back there hatched from the eggs laid a few days ago. Most of the burrow must be a madhouse of young, hungry mites jockeying for a way to get out and meanwhile eating everything in sight. Better to just forge ahead, the female must think, burrowing and laying eggs with the sort of tired resignation any mother of ten would recognize. It works for people; no reason it shouldn't work for our mites.

Even a creature with what seems to be so little by way of redeeming characteristics is not totally without benefit to its host. The young ones and the males, in their wandering around and poking into our follicles, generally police up the neighborhood by eating surface bacteria or the fungal spores that settle on us constantly. It's probably not sufficient to balance the damage their mother's burrows do to our skin, but it's only fair to mention it.

The most remarkable thing about mange mites is that we know relatively little about their personal habits and behavior. What may be exactly the same kind of creature inhabits the skin of a large number of mammals. Mammal skin, especially the dead outer surface, is pretty much the same from one species to another. A bear may not look much like a deer to us, but from the point of view of a mite, there is a lot of similarity. Mange mites are more or less serious in their effect, depending on the host. We could probably determine how long they have been living with a particular host on the basis of just how much damage they cause. Since no parasite benefits from seriously harming its host, over time natural selection produces parasites that are even less harmful. No parasite wants to be noticed; that's the road to being eliminated. Many parasites eventually live symbiotically with their hosts; a sort of you-scratch-my-back and I'll-scratch-yours relationship. The mange mites from dogs and cats can live on us temporarily but don't lay eggs or hatch young ones. Something is subtly wrong with our surface. It doesn't provide everything they need so they only survive and don't reproduce. What we really need if we are to ever understand the mange mites and even the diseases they sometimes cause is a sort of Jane Goodall for mites. Someone to watch individual mites carefully through a number of life cycles and note just what they do and when they do it. How do males and females find one another, and where? How do the females decide where to make their burrow? What do the young mites do before they are sexually mature? Exactly what and how much do they eat and where do they wander? Are they out mostly during the day or only at night? Where do they spend their leisure time?

All that is really needed is someone with sarcoptic (mange) mites on their wrist who would be willing to be watched, and someone to watch for a couple of weeks. The procedures for understanding animal behavior have been pretty well worked out in the last couple of decades; the procedures just haven't been applied to the creatures that live on the world that is us. We are pretty much still an unexplored continent with respect to what the mange mites do and when they do it. It's time we started to better know the territory and the tiny ones that live in and on it.

Face Mites or Follicle Mites

(*Demodex folliculorum* and *Demodex brevis*)

The human face is one of the least hospitable environments on the human landscape. It is mostly exposed to the elements, it usually bears little or no covering, and much of it is almost hairless and is washed frequently. It is for most of us in nearly constant motion. For the inhabitants it must be like living on a series of continually active geological faults, with floods and earthquakes an every-minute occurrence. In spite of this unhealthy set of circumstances, the face does have permanent inhabitants.

Follicle mites are found on practically all mammals and are universally present in and on some parts of the human skin. For a creature that lives so close to us so consistently, we know relatively little about the details of its lifestyle. Both species are too small to be seen without significant magnification, but we know they are there. The whole creature is about one-fourth the size of the period at the end of this sentence. Mites have four pairs of legs rather than the three pairs that characterize insects, and in the case of the follicle mites the legs are short and not very useful looking.

Most mites are nearly circular or spherical, but both species of face mites are long and thin, which fits them to their habitat in the narrow glands and hair follicles of the skin. They spend most of their lives head down in the cavelike atmosphere of our facial follicles.

I know you are wondering what they eat and how they live. Fortunately our follicle mites are less greedy and not quite as fast at reproducing as are those on less-favored hosts. *Demodex folliculorum* lives in hair follicles on the face and seldom ventures out on the surface. It can migrate from one follicle to another, but generally gets along fine munching on nearby skin cells and hanging out with a group of buddies at the base of one of your eyebrow hairs. When it gets too crowded, one or more will slowly and reluctantly move out, looking for a new home. If for some reason, such as a plugged pore, migration is not possible, crowding can get to be a problem. If there are more than about ten mites per pore, then what the dermatologists call eruptions may appear. This is as good a reason for washing your eyebrows regularly as I can think of. On the other hand (or rather in the other pore) is *Demodex brevis*, which as the name suggests is just a bit shorter than its companion species. It lives in the sebaceous or oil glands, which branch off the follicle like a secondary cave. *Brevis* is a loner, and your face offers a home to only one per gland as a rule. It is also most likely more mobile, wandering from gland to gland or even person to person when we are quiet and not rubbing our eyebrows in consternation.

Since no creature lives forever, the follicle mites must be reproducing right there on our face. They lay eggs that hatch into miniature versions of the adults. The young ones live in the same places as the adults. Questions that remain unanswered: How does a male follicle mite find a female? Is there mating going on right there in the follicle or do the mitish lovers seek out a shady nook on the surface under a clump of what must pass for trees if you are a mite?

One mystery has been solved. When microscopists first looked seriously at face mites (around 1900), some suggested that they appeared not to have an anus. This anatomical omission seemed so unlikely that genuine anal absence was not verified until 1977. Presumably they die of terminal constipation before they get very old. If you give the situation some thought, you should be overjoyed that our face mites have this particular digestive anomaly. Otherwise some of our favorite and more scatological insults would become literally true for everyone and would lose their significance.

Some have seriously suggested that it would be better for people generally if they did not know that mites live, eat, and breed in their follicles and on their faces. On the theory that knowledge is superior to ignorance and that if we can adjust to nuclear bombs we surely should be able to live with the understanding that a few mostly innocuous creatures share our skin, I submit the previous in pursuit of the self-knowledge that is the key to successful living. Socrates would want me to. And even if we wanted to get rid of the mites, we could not do so without harming ourselves much more than the mites harm us. Even shaving your eyebrows won't help; the follicles are still there. As long as there is a home and food, we can expect our follicle mites to remain in relatively harmless residence.

Chapter 5

WAY TOO SMALL TO SEE

There are fungi (molds) that live on everything, and we are no exception. Bacteria are everywhere there is food, and we are food. We cannot know they are there unless they irritate us, and they eat so little it doesn't often matter whether they are there or not. Fungi and bacteria are called microorganisms and, as the name suggests, they are smaller by an order of magnitude than any other ectoparasites we have considered. What that small size means is that they can live in very small habitats and can be present in very large numbers. To a fungus, the next finger over is the equivalent of another continent. The fungi called dermatophytes (skin plants) live mostly on the dead skin cells and dead structures such as hair and nails that are found all over the human landscape, and our resident bacteria live mostly on the sweat, fats, and oils secreted from the skin. Like nearly all small organisms, they must have water or at least high humidity if they are to survive or thrive. That means we should expect them to be at their best in the more sheltered and covered neighborhoods of the human landscape. When we began wearing clothes and shoes we made a whole array of new, warm, high-humidity tropical habitats for our resident microorganisms.

Fungi and bacteria do not eat in the same way that all the other creatures treated in this book do. Microorganisms soak up food from their surroundings through all their surfaces, and they have a lot of surface. For something as small as a single bacterium, the ratio of surface to volume is higher than it is for anything larger, and fungi have a similar ratio. That means that both can soak up relatively enormous quantities of food and can grow and reproduce more rapidly than anything else that lives. A bacterial cell can duplicate itself in twenty minutes if it is in the middle of enough food. That speed is directly a function of its small size. In general, the more food, the more reproduction and the larger the number of bacteria, which explains the smell of both babies and nursing homes. As we are less able to worry about what spills on our surfaces, the bacteria that can use it as food increase in numbers.

Fungi and bacteria grow in very different form than other creatures. Fungi forms long tubular strands of life, like the kind you see growing up from that bread slice that has been in the refrigerator too long. The strands can grow all around and between anything potentially useful as food, and can send even smaller strands into the interior of the smallest packages of potential nutrition. Fungal strands can completely cover the outside and fill the inside of your discarded cells or even the cells still attached to you. Fungi can grow in them like the blue mold in a chunk of Roquefort cheese. Bacteria grow as small separate units (cells) but the units are really only strands like those of the fungi divided up, and they too can fill any source of food with their growth in a matter of hours.

Fungi and bacteria are themselves food for many other living things. When we eat mushrooms, cheese, or even sauerkraut, we are using fungi and bacteria for food. The presence of fungi or bacteria in those foods makes them more nutritious. An active culture of fungi in our discarded skin cells may make them more complete nutrition for the mites and others that eat them in the same way that the active bacterial cultures in yogurt make it a more complete food for us than was the milk from which it was made.

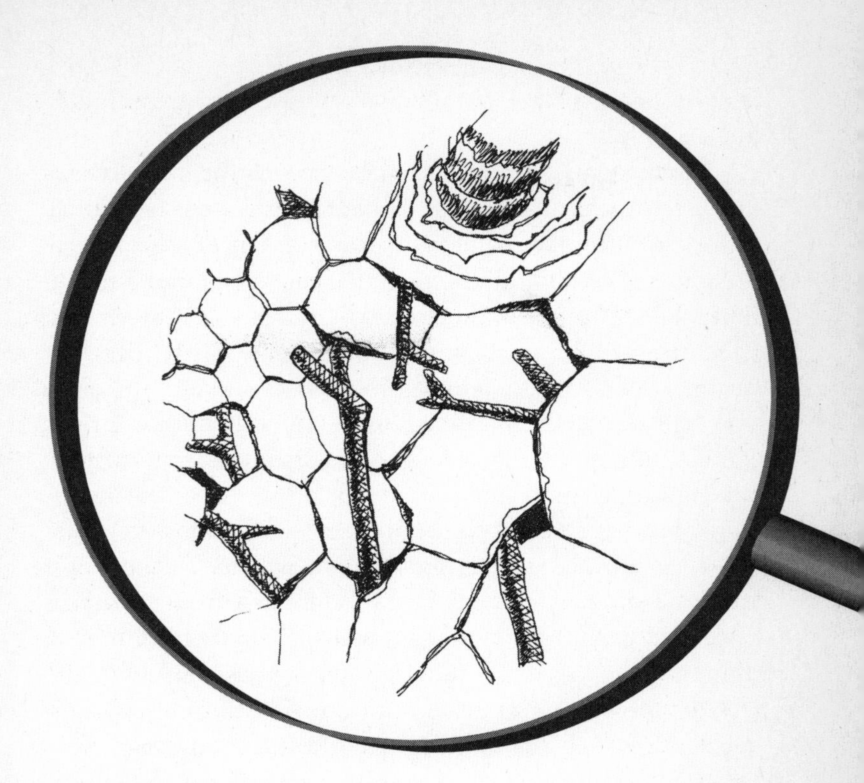

Dermatophytes

(*Trichophyton* and *Microsporum* species)

Each of us walks through life scattering leftovers for other creatures who can use them. Not just the fast food fragments that we discard, but microscopic bits of us that are like manna to many of the molds. We shed hundreds of hairs of all sizes and thousands of skin cells each day. When we are indoors we treat all this dust as a problem and spend a lot of energy vacuuming and sweeping it up. When we are outdoors all that leftover dead stuff falls to the ground and becomes a resource rather than a disposal problem. What on a dry floor is just a fossilized skin cell becomes on a damp soil surface a tasty food item for whatever can find it and use it. Since this daily distribution of food has been going on for millions of years, the soil has a large number of organisms that specialize in the effective recycling of our excess, discarded body parts.

Farmers and gardeners are not the only people close to the soil. Most of us come in daily contact with airborne soil dust that a few days or hours ago was a part of the earth. If the soil dust has in it those fungi that can live on our hairs or skin cells and use them for food, when that soil contacts our skin, fungi, who don't know much, cannot distinguish between dead hairs or skin cells on the ground and the dead skin cells or hairs still attached to us. If the humidity is high enough, like in the sheltered nooks, crannies, and corners of our landscape, those fungi live right on us. We don't know that it has happened, because they cause us almost no trouble. The skin and hair they are living on is in turn shed after a few days, and new dust with new fungi move in to take their place as a part of the walking, talking community of organisms that we are. If the fungi get too aggressive and begin growing too fast, they irritate our skin and our skin responds by shedding cells just a bit faster, and the overly aggressive fungi are gone. We are likely to be aware of our microflora (moldy residents) only in those places on us where skin cells are shed less easily, or if we become somehow less capable of making new ones. Then the doctor may tell us we have athlete's foot or *tinea pedis*. That means a fungus is living on the skin between our toes where old cells hang on for longer and where shoes keep the humidity high. Most of our skin surface is less hospitable to the fungus, not only because it is dryer but because our pores make and ooze out onto the surface chemicals that actively inhibit the growth of fungi. Our feet and hands have fewer pores and less capacity to inhibit. Most of the time we and our fungi live in a delicate and persistent balance, where we shed the parts they are using at just the right rate so that both we and the fungi do well.

What is commonly called ringworm has been known as a disease of the human skin for centuries, but until the nineteenth century the cause was obscure. As the name suggests, it was thought to be caused by vermin of some sort. Now we know it's cause by fungi that are able to live on our skin and hair and that are present on nearly everyone. The disease appears when the fungus grows a bit too fast, we respond a bit too slowly, or we develop an allergic reaction to whatever the fungus is producing as byproducts of its reckless, over-fast life on our surface. Our skin and hair don't look much alike to us, but they are very similar chemically, and a fungus will live on anything it can use as food. The ringworm infections commonly found in infants and children are often in the scalp and hair, probably at least in part because we tend to keep those parts covered.

There is a subtle variability in the fungi that cause ringworm, and one of the reasons they persist in humans is that we spend a lot of time together, often in close quarters. Should a variant of some dermatophyte have success at maintaining that delicate balance between growing too fast and causing a self-limiting inflammation (speeding up cell shedding) or growing too slowly and never getting established before the skin or hair is shed, it will find that getting from one source of food to another isn't much of a problem. Many of the dermatophytes have given up the usual means of dispersal for fungi (tiny spores that blow around in the wind) and depend on our social habits for their existence. They travel on flakes of skin and pieces of hair that we leave behind on beds, chairs, or clothes. We seem to have domesticated our fungi so they live pretty much in harmony with us. What the fungus gets from such a relationship is good nutrition, a safe place to live, numerous nearby hosts, and a convenient travel plan. What we get is limited to our not having an obvious infection or disease most of the time.

You may conclude that the fungus certainly has the better of the bargain, but if it's going to live on us anyway, better to have an accommodating fungus than one that is less pleased with its environment and causes trouble all the time.

We have made the fungus into a more benign creature by shedding those that grew too fast or irritated too much. Our response to it (an unconscious one) has helped to make us into a suitable host for one of our most persistent guests. As long as the fungus doesn't irritate too much, we don't shed it and life on earth is richer and more abundant. When talking about the balance of nature we usually think of wolves and moose or grass and gazelles, but dermatophytes and our skin are at least as enlightening and useful an example, and one that is much closer to home for nearly all of us.

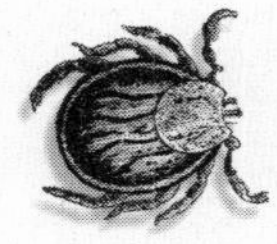

Hair Fungi *(Trichosporon beigelii* and others)

While the ringworm fungi can live on hair and invade the dead skin as well, there are at least a few of their moldy relatives that specialize in hair and never live anywhere else on us. They too live in the soil, and are found on nearly all of us nearly all the time, but only occasionally cause noticeable symptoms. Those symptoms involve no discomfort or disease, unless split ends make you uncomfortable or brittle hair is thought of as a disease.

If you are the size of a fungus, a single hair is a large and complicated thing. Hair has a tough outer cuticle and an inner core. If the cuticle is damaged, the hair fungi can get a foothold on the tender interior and live kind of like a beetle burrowing under the bark of a tree. The infected hair develops a colored clump at the point where the mold has taken hold, and the hair itself becomes brittle and is likely to break off. The abundant presence of hair fungi does seem to be related to somewhat lax habits of hygiene, or having your hairy parts close to the soil, where the fungi normally live. They do best in places where the hair is warm and damp; you know where those places are. When not on our hair, the fungi can live in soil, stagnant water, or spoiled food. All of which means they are near us all the time, and clean, healthy hair is your only defense.

Some hair fungi produce white lumps on the hair, and others black. We could probably utilize genetic engineering to make other colors. Their potential as items of personal hair fashion has not been exploited yet, but being the first brunette with just-visible white flecks a permanent part of your coif might be appealing to someone in pursuit of an individual fashion statement. If we can just get comfortable with the idea of a mold growing on our hair, someone will develop a whole set of franchised salons where you can get your hair invaded with a fungus of an appropriate and becoming color.

Tooth Amoebas *(Entameba gingivalis)*

We all looked at amoebas in biology class. They live in stagnant water and ooze around from place to place. What's the closest thing to stagnant water that's near us? We have all seen the TV commercials about morning mouth. That's stagnant water in there, folks, and we shouldn't be at all surprised that at least one amoeba lives in it. Brushing won't get rid of them. The mouth has a lot of very small places where a microscopic amoeba that can change its shape will be able to avoid even the most diligent campaign.

Should you worry about what your amoeba is doing while you are asleep? Alfred E. Neumann wouldn't worry, and neither should you. The question you need to ask is, what do amoebas, or this particular amoeba, eat? They eat bacteria; they eat tiny bits of leftover organic matter that are small enough for them to surround. We haven't studied the activities of this particular creature as much as might be desirable considering how close we are, but what we know suggests that they cause no diseases and probably help keep our mouths free of some less-desirable invaders. It's at home in there, lives in balance with us, and seems to be helping to keep other things in balance. Only if it reproduces a bit too fast and there are too many of them will we ever know they are there, and even that sort of population explosion is likely to be caused by some general decline in health. Hard as it is to say it, there is an amoeba that lives in our mouths in harmony with us, even if we don't sing. Welcome your amoebas, visualize them if you like, enjoy their presence insofar as you can: they are a permanent, healthy part of the neighborhood, and without a microscope you will never know they are there.

Bacteria

(Many species, including *Staphylococcus, Corynebacteria,* and *Micrococci*)

We come finally to the very smallest things living on us that are capable of growth and reproduction: the germs, or bacteria. They are small enough so that an uppercase O about the size of a bedbug could hold numbers of bacteria equal to the population of Omaha, Nebraska; and in just one layer. Small enough so that a hundred thousand of them laid end to end would just stretch across the palm of your hand. And if your palm was sweating (as it might be if you knew they were there), and if some bacterial food was available, as it might be, there could be at the end of an hour 800,000 of the little things forming a still-invisible line across your palm. The numbers of bacteria on one person's skin at any one time equal the numbers of people who have ever lived. We all know that germs cause disease, but we need to know more than that. Actually, a very few kinds of germs cause a few kinds of diseases. Nearly all the bacteria are either benign or beneficial, and that is indeed a good thing, since we provide them with a variety of different foods in different locations on our landscape, and there are a lot of different kinds that find us to be a suitable or even desirable home. For something the size of most bacteria, the groove beside your fingernail is a giant canyon that is warm and wet at the bottom and dryer and cooler at the top edge. Somewhere in that gradient of conditions, things are just right for living and growing.

Most of our growing germs are confined to a few locations on the human landscape, usually the wetter and warmer areas. Some potentially dangerous ones live only inside your nose, where they cause no problems of any kind. They will thrive in the grooves and creases around your fingernails, should they get there, and from there can be distributed nearly anywhere on your own or someone else's body. Nose pickers and fingernail chewers take heed. If your nose-dwelling bacteria get to places where there is broken skin or a small wound, they can cause serious infections.

Washing and showering has little effect on the number of bacteria that live on you, although you can move them around from place to place. The kind of soap you use has likewise little effect on the numbers or locations of your resident germs. How many we have depends mostly on the kinds and amounts of food our skin provides, and the kinds of environments we make for them on our surfaces. If you are a messy eater, you probably have a lot more surface bacteria than other people. When we are either very young or very old, we usually host more of our tiniest residents. Clothes provide a warmer and damper environment for anything that lives on our skin, including the bacteria, and our surface flora probably changes with changes in fashion. Tighter clothes make for a different surface climate than looser ones, and no clothes at all exposes us and our germs to the ultraviolet rays in sunlight. These are the rays that cause sunburn and kill bacteria. Enough exposure to ultraviolet light to reduce your covering of bacteria by much would carry a great threat of skin damage from severe sunburn. Better to let the little critters live there than to try finding ways of getting rid of them.

Do our naturally occurring surface germs do us any good? By using up some of the food on our skin (the oils and minerals in sweat and such) they make it more difficult for other nonresidents to get a foothold. Our residents do contribute to our individual and unique odors, by manufacturing their own chemicals, and automatically provide a significant part of our personality as perceived by others. Since our skins are each subtly different, and harbor slightly different combinations of surface bacteria, you should probably do what you can to keep the ones you have and possibly even cheer them on just a bit. You and they have been living together for decades (longer than most other relationships). Such faithfulness deserves at least the minimum reward of recognition and continuity.

Our bacteria are the most numerous and least well-known of the residents we generously provide for. They are just too small to follow individually by any of the mechanisms available to investigators, and since an individual can become a crowd in a few hours, where the individuals are is significant. They reproduce by simply dividing themselves in two, which means reproduction doesn't require a partner. They can become in an instant something slightly different from what they were. Mutations—changes that make a slightly different organism—occur with greater frequency in bacteria than in any other sort of creature. Changing the germy equivalent of hair color or skin pigment is the work of seconds, and all future offspring of that mutation will have the same change. You can't keep an eye on them, nor can anyone else, and whether you enjoy thinking about their presence or are repulsed by the thought of some bacteria on almost every surface and in every nook and cranny won't make any difference to them at all. They don't know much, but they know enough to recognize a good place to live when they find it, and being a good place to live is something we might all desire.

AFTERWORD

We and Our Beasts

While the preceding is far from a comprehensive account of our residents and companions, even a cursory look at those things that share our most intimate moments shows us how little we know about many of them. We know more about rabbit fleas than we do about human fleas, and nearly all we have learned about both fleas and lice has to do directly with their relationship to human diseases. Those that are too small to see easily are the most thoroughly neglected. We will continue to maintain deep gaps in our knowledge until we begin to look at these fascinating creatures as interesting in their own right, not just as something we should be getting rid of. Until we understand face mites in the same way that we have come to understand the mountain gorillas or the chimpanzees, we will miss a lot. Until scabies mites are viewed as an intriguing population living on some of the most interesting real estate in the world, we probably won't know enough to control their numbers as well as we could. As long as we can't see the fungi that we feed and shed so frequently as being interesting for themselves alone, we will have difficulty understanding how they can occasionally cause disease for some of us and live in permanent comfort with others. As long as many of us think that killing all our bacteria would be a good idea, we will never understand the more complicated ideas of living in harmony with other easier-to-see creatures. Furtive fauna are fantastic; we might as well be fans.

Further Reading

Alexander, John O'donel. *Arthropods and Human Skin*. New York: Springer-Verlag, 1984.

Barnes, Robert D. *Invertebrate Zoology*. Philadelphia, PA: W. B. Saunders, 1974.

Belding, David L. *Textbook of Clinical Parasitology*. New York: Appleton, Century-Crofts, 152.

Busvine, J. R. *Insects, Hygiene and History*. London: The Althone Press of the University of London, 1952.

Cameron, Thomas W. M. *Parasites and Parasitisms*. New York: John Wiley and Sons, 1958.

Cheng, T. S. *General Parasitology*. San Diego, CA: Academic Press, 1974. 2nd edition, 1986.

Harwood, R. F., and M. T. Jones. *Entomology in Human and Animal Health*. New York: Macmillan, 1979.

Lehane, Brendan. *The Compleat Flea*. New York: Viking Press, 1969.

Marples, Mary J. *The Ecology of the Human Skin*. Springfield, IL: Charles C. Thomas, 1965.

Nutting, W. B., and H. Beerman, "Demodicosis and Symbiophobia: Status, Terminology and Treatment," *International Journal of Dermatology* 22 (1983): 13.

Oldroyd, H. *The Natural History of Flies*. New York: W. W. Norton, 1964.

Parish, L. C., W. B. Nutting, and R. Schwartzman, eds. *Cutaneous Infestations of Man and Animal*. Westport, CT: Praeger Publishers, 1983.

Riley, William A. *Introduction to the Study of Animal Parasites and Parasitism*. Ann Arbor, MI: Edwards Brothers, 1939.

Rippon, J. W. *Medical Mycology*, Third Edition, Philadelphia, PA: W. B. Saunders, 1988.

Rosebury, Theodor. *Microorganisms Indigenous to Man*. New York: McGraw-Hill, 1962.

Rothman, Stephen, ed. *The Human Integument*. American Association for the Advancement of Science Publication 54 (1959).Rothschild, Miriam, Y. Schlein, K. Parker, C. Neville, and S. Sternberg, "The Flying Leap of the Flea." *Scientific American* (November 1973): Vol 229, No. 5, pp 92-100.

Trager, William. *Living Together. The Biology of Animal Parasites.* New York: Plenum Press, 1986.

Wilde, J. K. H., ed. *Tick-borne Diseases and Their Vectors*. Edinburgh: University of Edinburgh, Center for Tropical Veterinary Medicine, 1978.

Wooley, Tyler A. *Acarology: Mites and Human Welfare*. New York: John Wiley and Sons, 1988.